The Cuckoo

THE UNINVITED GUEST

FOR JANA AND VIKTOR

OLDŘICH MIKULICA | TOMÁŠ GRIM | KARL SCHULZE-HAGEN | BÅRD G. STOKKE

FOREWORD BY NICK DAVIES

The Cuckoo

THE UNINVITED GUEST

Table of contents

A portrait of a brood parasite and its hosts

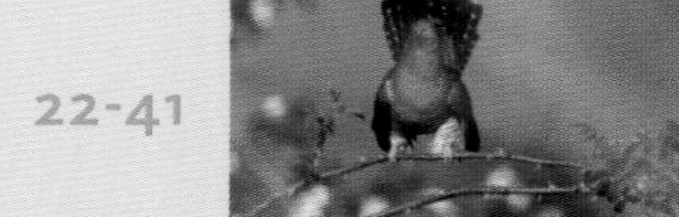

Cuckoo eggs

Cuckoo chicks

The complex co-evolution of brood parasitism – theories and research

Cuckoos in a changing world

Passion for a parasite

1. Foreword

This is a book once opened that is impossible to put down. There is delight and astonishment on every page as Oldřich Mikulica reveals, in photographs of unparalleled detail and extraordinary artistry, the secret life of Nature's most notorious cheat. They are accompanied by an authoritative text from a distinguished international team of researchers, which brings us right up to date with the latest discoveries, and so the book is also a feast for the mind, a celebration of how wonderful our natural world can be.

The cuckoo is a familiar harbinger of spring right across Eurasia, from Western Europe to Japan, but although many will immediately recognise the male's distinctive two note call – "cuck-oo"- far fewer have seen this elusive bird. "Shall I call thee bird, or but a wandering voice?" wrote the English poet, William Wordsworth. This wandering voice has inspired more myths than any other bird, and has been supposed to predict human fortune in life and love. But the cuckoo's true lifestyle is stranger than any myth, for it lays its eggs in the nests of other species and tricks them into raising a cuckoo chick rather than a brood of their own.

Marvel then, in the pages here, at the extraordinary sight of a little warbler feeding a cuckoo chick and seeming to risk being devoured itself as it bows deep into the enormous gape to feed a young bird many times its own size. How does the cuckoo get away with such outrageous behaviour? It used to be thought that the cuckoo's lack of parental care was a result of defective design by a Creator, and that other species were only too pleased to give the poor cuckoo a helping hand. Ever since Darwin, we have realised that the cuckoo's parasitic habits have evolved by natural selection and that the hosts, in response, have evolved defences. The result has been a continuing evolutionary arms race between cuckoos and hosts, involving adaptations and counter-adaptations.

Some of the most remarkable photographs here show this battle in action: the drama of a pair of great reed warblers attacking a female cuckoo as she tries to lay an egg in their nest and that of a warbler puncturing and ejecting a cuckoo egg. But cuckoos have evolved speed and secrecy to enable them to slip past host defences, so marvel, too, at the cuckoo's mimetic egg, which is often a good match of the background colour and spotting pattern of the host eggs. If the hosts are fooled by the cuckoo egg then the worst is still to come, for they are now sitting on a bomb set to explode in eleven days' time, one that will destroy their clutch. Even those familiar with what happens next will be amazed by Mikulica's photographs of the newly-hatched cuckoo, still naked and blind, ejecting the host's eggs and chicks from the nest one by one and, perhaps more extraordinary still, those of the host parents calmly watching the destruction of their own brood, yet doing nothing to interfere.

For me, two of the most poignant images appear in the final chapter. One might have been the easiest of all to take, that of Mikulica's hide, but it is a reminder of the extraordinary patience and skill that lie behind every picture. It's a privilege to eavesdrop on what he has seen, hard won and fleeting moments that take just a few seconds, such as cuckoo egg laying and host egg rejection. The other is of his grandson, Viktor, gently holding a young cuckoo. A few weeks later, this bird would have attempted a flight south to its African winter quarters, enduring a non-stop flight of some fifty to sixty hours over the Sahara desert. Satellite tracking reveals that the alarming decline in cuckoos in some parts of Europe is likely to reflect increasing droughts and tougher feeding conditions on migration. We should count ourselves fortunate that cuckoos continue to visit our increasingly impoverished natural landscape. Their decline is a potent symbol of our diminishing natural world, and this wonderful book is a stark reminder of how sad it would be to deprive future generations not only of our harbinger of spring, but also of some of the most extraordinary natural history on earth.

Nick Davies

Male cuckoo

Female cuckoo

A portrait of a brood parasite and its hosts

2. Discovering a very special bird

The cuckoo is a special bird – and an immensely popular one. Everyone knows its call and every child can mimic the distinctive ‘cuckoo – cuckoo’ sound. This call, as striking as it is simple, is synonymous with spring, heralding the season of sunshine and warmth like a peal of bells. However, how many people have actually seen this harbinger of spring? Compared with other familiar summer favourites, such as the barn swallow or swift – not that many. Cuckoos can be very secretive birds, generally hiding from view as part of their breeding strategy, and when we see one at all it is usually in flight. Although it is general knowledge that cuckoos lay their eggs in the nests of other birds, this is rarely nowadays explained to young people in school. More detailed knowledge of the unusual reproductive strategy of brood parasitism is mostly reserved for the keenest of naturalists. In fact, cuckoos are popular among naturalists largely because of this brood parasitic habit. After all, the cuckoo is the only full-time brood parasite among Europe’s birds, apart from its distant southern European cousin, the great spotted cuckoo. Cuckoos are not related to any of the host species that they parasitize and which raise their young and the hosts of ‘our’ cuckoo are in fact all songbirds (see chapter 4).

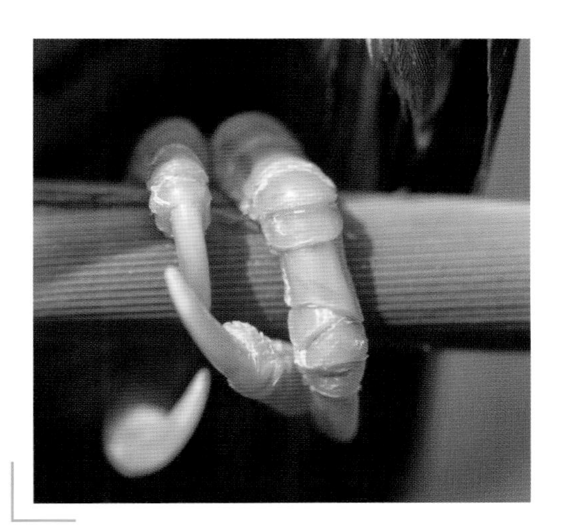

Zygodactyly, where two toes face forward and two back, is a peculiar trait shared between cuckoos and some other climbing birds such as parrots and woodpeckers.

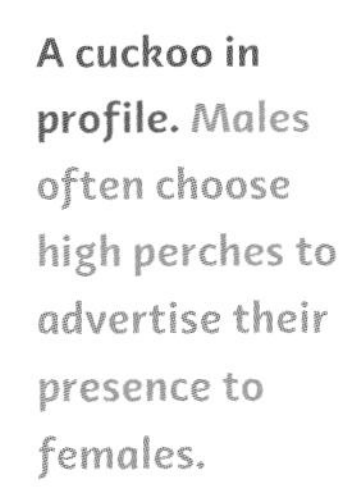

A cuckoo in profile. Males often choose high perches to advertise their presence to females.

Let’s start with a short overview of cuckoo biology. ‘Our’ cuckoo, officially the Common Cuckoo, *Cuculus canorus*, and its relatives have their own category in the scientific list of bird groups, the order Cuculiformes. This contains almost 150 species of cuckoos, but in fact only about 50 of them are brood parasites. Many cuckoo species build their own nests, incubate their own eggs, and raise their own young. Of all the characters that define members of the Cuckoo order, one that is only known by true specialists is the curious shape of the foot. Cuckoos (in common with some other groups of climbing birds, such as woodpeckers) are peculiar in that, of the four toes on each foot, the two outer ones point backwards while the two inner ones point forwards – an arrangement known to zoologists as zygodactyly.

The common cuckoo has a body 33 cm long and is about the same size as a collared or turtle dove, but weighs 90–130 g, slightly less than a turtle dove. It is a fast and skilful flier, with long pointed wings, and can cover very long distances. In flight it looks much like a bird of prey, such as a sparrowhawk, especially with its barred underside. The males are mainly slate-grey above, but females are more variable with differing proportions of brown tones and some females are completely reddish brown (see chapter 5).

The aerial acrobat. Cuckoos are skilful fliers.

A habitat generalist. Cuckoos can be found in many different kinds of habitat, and reedbeds surrounded by tall trees are a favourite in some areas.

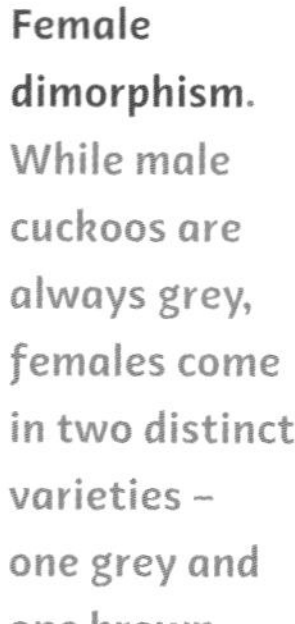

Female dimorphism. While male cuckoos are always grey, females come in two distinct varieties – one grey and one brown.

The cuckoo has become rare in many places today, due to substantial habitat change and loss. But 30 years ago it was an everyday bird, common everywhere for three main reasons: its very widespread distribution; its broad habitat spectrum; and its 'utilisation' of many different host species (see chapter 4). The common cuckoo breeds throughout Eurasia, from the Atlantic to the Pacific, occupying two entire continents, and within this area is absent only from Iceland and the subpolar regions. Hardly any other European bird has such a broad ecological niche as the cuckoo – it is a habitat generalist, whose own requirements match those of its many host species. Woodlands and semi-open landscapes are favoured when searching for food, but for egg-laying practically any landscape type is used, from the grasslands of the subalpine zone to moors, heaths, reedbeds, arable land, and saltmarshes elsewhere.

Cuckoos feed chiefly on large insects, including beetles, dragonflies, and grasshoppers. Caterpillars form an important part of the diet, particularly the hairy caterpillars of several moth species, which occasionally occur in very large numbers on their food plants but are ignored by the majority of other bird species. Whenever a periodic infestation of processionary moth caterpillars occurs in pine or oak woods, cuckoos will travel to gorge themselves at the feast and in the 20th century it was still possible to encounter up to a hundred cuckoos foraging simultaneously. Today however, the increasing rarity of both cuckoos and moths have rendered such scenes a thing of the past (see chapter 14). Cuckoos are occasionally seen on the ground in wetlands, hunting for small frogs, and female cuckoos regularly consume birds' eggs and even nestlings.

A peculiar diet. Juicy, poisonous and hairy caterpillars are a favourite food of adult cuckoos.

Cuckoos are migratory birds, arriving on their European breeding grounds from mid-April. As early as June, the first adults leave the breeding site to migrate again through southern Europe and North Africa to their wintering areas in and around the tropical rainforests of Central Africa, where they moult their feathers. Young birds leave later, and are occasionally still encountered in their wider breeding range in August, or even September. 'Our' cuckoos are truly nomadic – spending a lifetime on the move, interrupted only by short periods loitering in their breeding or moulting quarters. Using today's satellite transmitters, individual cuckoos can be followed along their complete migration route (see chapter 14).

As soon as the males have arrived on the breeding grounds, they declare their territory with their striking, well-known call, often singing continuously for several minutes at a time. At the end of the breeding season, in June or early July, they stop calling, remaining silent until the following spring. Females have a very different voice, though one rarely hears their trilling, giggling sound, rather reminiscent of the bubbling call of the little grebe. It is usually heard immediately following egg laying.

Male and female cuckoos form no close bond, and have a promiscuous mating system. Their breeding habits relieve them of the task of caring for their young, a job left to the host species as the cuckoo's egg smuggles a changeling into their nest. The advantage for the parasite is a disadvantage for its victims, the host birds. If the young cuckoo is to grow and prosper, the

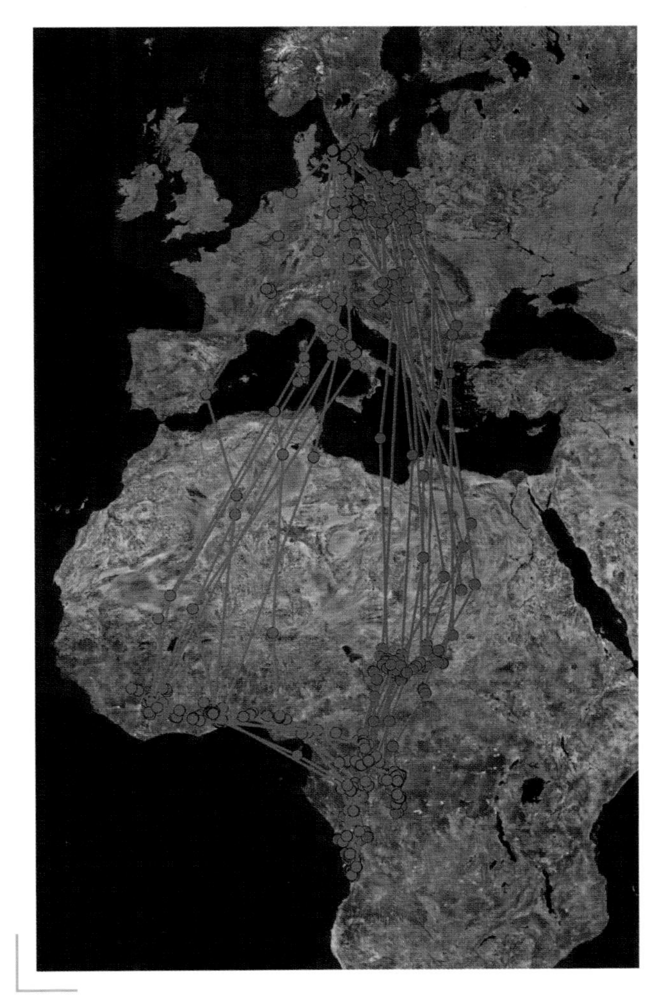

Cuckoos are long-distance migrants. Autumn (red) and spring (blue) migration of Danish and Swedish cuckoos as disclosed by satellite tracking of birds equipped with transmitters (Illustration by Mikkel Willemoes Kristensen and Kasper Thorup).

A male cuckoo chases a female. Cuckoos are promiscuous and never establish permanent pair bonds.

host birds must lose their own brood (see chapter 7). These opposing interests result in the – now famous – evolutionary arms race between brood parasites and host birds. This forms an escalating spiral of tricks and countertricks, of adaptations and counter-adaptations that over the course of evolutionary time have led to the highly sophisticated cuckoo strategies that we see today.

In the chapters that follow, we will take you on a journey into the life stories of the common cuckoo and reveal some of their mysteries through the spectacular photographs of Oldřich Mikulica, which give us the opportunity to peer into the hidden aspects of the cuckoo's life. Pictures often say so much more than words and are given pride of place in this book. They take us on a journey through the breeding season of this parasite and its hosts, with a special focus on a particular reedbed in southern Moravia, Czech Republic (see chapter 15). Our text is built around these photographs and presents an overview of our current state of knowledge (often very different from what we 'knew' just a couple of years ago), and of the theories regarding the behavioral strategies evolved by avian brood parasites – and the counter-adaptations of their hosts.

3. An unconventional family life – Brood parasitic strategies around the World

Most bird species live a family life similar to that of humans. They form pairs, make a home for themselves and bring up their children, expending lots of energy in the process. However, about one percent of the world's 10,000 or so bird species have a radically different approach to the idea of a family. They do not build their own nests, they do not incubate their own eggs and they do not raise their own children. In fact, their only close encounter with their offspring occurs while they are laying their eggs. These birds are the avian brood parasites. Instead of expending lots of time and energy raising their young, they leave all the parental responsibilities to a host species, using it as foster parents. Some 100 avian brood parasitic species are 'obligate', that is, they always lay their eggs in the nests of host species. Many others are 'facultative' brood parasites which sometimes lay eggs parasitically in the nests of both conspecifics and other species. In most of these species, parasitism is a rare event – a parasitic female lays part of her clutch into other nests but then builds her own nest and lays the rest of her clutch there and incubates it normally. Coots, moorhens, ducks and starlings, among others, behave in this way. In the wider

natural world, brood parasitism is not restricted to birds and can also be found in insects, amphibians and fish.

At first sight, it appears that a life free from all parental duties represents a luxury for the parasite and one that we should expect to find more often. However, brood parasitism is a complex strategy and has evolved independently only seven times in diverse groups of birds: cuckoos (on three independent occasions), cowbirds, honeyguides, parasitic finches and, intriguingly, in a single species of duck. While all are obligate brood parasites, relying completely on other bird species to raise their chicks, they vary considerably in how this strategy is realised. The number of species used as hosts differ – specialists exploit a single host species, while generalists use many different host species. The eviction behaviour of their young differ – some grow up in the nest alongside the host chicks, while others evict all of the host eggs and chicks. They show various degrees of egg and chick mimicry – some lay eggs similar to those of their host and produce chicks that closely resemble their host chicks, while others show only one or neither of these adaptations.

Because of the damage that brood parasitism inflicts on the breeding effort of the hosts, hosts have evolved an array of defences against parasite attack. These can be roughly divided into 'frontline' defences (that is, prior to

A selection of avian brood parasites (·) **brown-headed cowbirds; female** (left) **and male** (right) (Photo: James Rivers), (··) **a female parasitic cuckoo finch** (left) **and its red bishop host** (right) (Photo: Claire Spottiswoode), (···) **a greater honeyguide** (Photo: Claire Spottiswoode), (····) **black-headed ducks; female** (left) **and male** (right) **with their red-gartered coot host** (centre) (Photo: Bruce Lyon), (·····) **a juvenile female common koel** (Photo: Mohammed Mostafa Feeroz).

the cuckoo laying its egg), and adaptations at the egg, chick and fledgling stages. Brood parasites use various gimmicks to deceive the hosts into slaving for them which are matched by the hosts' manoeuvres to rebuff the parasites. After all, these are two sides of the same coin – without host defences there would be no need for parasites to mimic the patterns of the host eggs or to be so secretive when laying their eggs. And without these tricks there would be no need for host defences to be so fine-tuned.

Of the avian brood parasites, the cuckoo family (Cuculidae) contains the greatest number of individual species. We know very little about the biology of most of them and much of our knowledge of the life of cuckoos stems from studies of the common cuckoo and the great spotted cuckoo, which is principally an African bird, but also breeds in parts of southern Europe.

We sometimes imagine that cuckoos enjoy a lazy life – without parental duties and with lots of free time – but this is a naive and one-sided view. Indeed, while cuckoos do not invest in nest-building, incubation and chick-rearing, they do incur various costs not paid by non-parasitic birds that behave 'decently'. Reed warblers, white wagtails or robins do not expend time and energy searching for host nests, and then watching them to spot the right time for laying; they do not risk being attacked, or even killed, by their hosts; they do not lose their precious eggs when they are rejected by the host; and their non-parasitic embryos do not pay the extra costs of hatching from the strong eggshells typical of cuckoos. Although it is hard to quantify these costs, we can be sure, contrary to traditional ideas, that cuckoos and other nest parasites do not escape the costs of breeding – they simply exchange the traditional parental costs for alternative, parasitic costs. Cuckoos are not lazy birds at all – in fact they are kept pretty busy making sure that their progeny survive and prosper. Let us now take a closer look at how they manage it.

Two cuckoo eggs and one, smaller, reed warbler host egg. The cuckoo eggs were laid by two separate females, as is evident from different colour and patterns on the two eggs. Such cases are rare, due to territoriality among female cuckoos.

Female cuckoos are territorial during the egg laying period and do not tolerate others on their claim. Sometimes other females trespass, especially in areas with high densities of cuckoos and hosts, and two cuckoo eggs can occasionally be found in the same nest when the territory holder and an intruder both lay in it. Parasitized nests containing more than one cuckoo egg only occur regularly in host populations experiencing extraordinarily high parasitism rates – in reedbeds in Hungary more than half of the great reed warbler nests contained one or more cuckoo eggs, with a record six eggs in a single nest. Female cuckoos lay one egg every second day, and can lay more than 20 eggs in a season, provided that conditions are optimal and sufficient host nests are available. However, most females lay fewer eggs, around 10 per season. Research suggests that some females without their own territory, perhaps young and inexperienced ones, may not lay any eggs in a season, or only a handful. Perhaps they are just waiting for an opportunity to take over a territory? Like the youngsters of many species, including humans, they may need to gain experience before becoming successful.

4. Finding foster parents – Host selection

While male cuckoos spend most of their time displaying to females, mating, and chasing off other males, females spend most of their time during the breeding season searching for host nests. A female will sit motionless for hours high up in a bush or tree, or perhaps even on a telegraph wire, watching the activity of potential hosts. Perches are selected to provide a good overview of the area so that she can monitor the whereabouts of host nests and, not surprisingly, hosts nesting close to such perches are more likely to be parasitized than those nesting further away.

Exquisite egg mimicry. A marsh warbler nest with one cuckoo and three host eggs (Photo: Per-Harald Olsen).

In the Western Palearctic, cuckoo eggs or chicks have been found in the nests of more than 150 species, including virtually all of the passerines breeding in forests, meadows or marshes, giving the impression that female cuckoos lay their eggs haphazardly in any nest that they encounter. The truth could not be more different. Although the list of species found hosting cuckoo eggs or chicks is long, only about 40 species are regularly used as hosts and the remainder of the list comprises species which have only been accidentally parasitized a few times.

Male cuckoos quarrelling. Males compete over the best sites for attracting females.

Each female cuckoo belongs to a host-specific tribe – called 'gens' by specialists – often mimicking the eggs of her particular host both in ground colour and spotting pattern. The spatial distributions of the cuckoo tribes found in Europe vary substantially. The meadow pipit is the favourite host in the mountains of Scandinavia and in northern European heathlands, while bramblings and common redstarts are cuckoo favourites in the northern boreal forests. Great reed warblers, rufous-tailed scrub-robins, Orphean warblers and woodchat shrikes are commonly parasitized in Southern Europe. Other cuckoo tribes, for instance those utilizing reed warblers, robins, white wagtails and wrens are distributed over larger areas. Some species are used as hosts very locally, like the great grey shrike in Alsace, the chaffinch in the Moscow area and the chiffchaff in Bavaria.

It has long been known that inheritance of egg colour is passed down through the maternal line, that is, from mother to daughter, without any influence of the father's genes. Hence, strictly speaking only females belong to a specific tribe, while males can mate freely with females of any tribe without disrupting the inheritance of egg mimicry. Cuckoos can therefore maintain female races specific to particular hosts while still remaining a single species. Successful parasitism relies on host specificity so what cuckoo host selection strategies could maintain specialisations such as host egg mimicry?

Many birds remain faithful to the place where they were born. This 'natal philopatry' tends to bring female cuckoos into contact with the same host species in whose nest they were reared. But this effect is only pronounced in very homogeneous habitats where few potential hosts breed. A similar effect would be achieved if females became imprinted on the habitat type in which they were reared – subsequently seeking out the same habitat in which to lay their eggs as an adult, even in geographically distant areas. The reed warbler is the dominant species in many reedbeds, with very few or no other potential hosts present. If, after wintering in Africa, female cuckoos habitually returned to the very same (by philopatry) or a different reedbed (by habitat imprinting), most or all of the nests encountered would be of the correct host. While this illustrates the point, it is an extreme example – most habitats are much more heterogeneous than reedbeds and consequently contain a number of different potential host species. In such habitats, nest-site choice may provide another means of maintaining egg mimicry. Cuckoos could search for nests placed in particular locations such as low bushes or the eroded root systems of forest trees, and parasitize all the nests in these and similar places. However, because several potential host species with clearly different egg colours may build their nests in similar sites this host selection strategy is prone to error. Could there be alternative host selection strategies to ensure that the cuckoo egg is laid in the 'correct' species nest? The answer is yes.

Radio-tracking and genetic analyses have confirmed that each female cuckoo shows a strong preference for a particular host species. Many studies of different hosts and different habitats have shown that each

Tracking cuckoos. A cuckoo fitted with a radio transmitter. Radio telemetry provides a convenient tool for tracking cuckoo movements within its breeding area.

female cuckoo parasitizes only a single host and avoids alternative hosts available in the same habitat. This host preference explains why different host species with similar habitat and nest-site preferences are parasitized with suitably mimetic eggs even though each hosts' eggs are different. We still do not know exactly how this host preference works. A plausible explanation is that each female cuckoo selects the host species that she grew up with, having become imprinted on its appearance or vocalisations while growing up in its nest.

It is theoretically possible that host selection acts at even more subtle levels than just recognition of the correct species. Individual female cuckoos (and hosts) lay eggs that remain remarkably similar in colour and patterning throughout their lives. Furthermore, different female cuckoos belonging to the same tribe lay eggs that are more similar to each other than to those of females of another tribe. So, eggs of females belonging to the meadow pipit cuckoo tribe are clearly different from eggs belonging to the garden warbler cuckoo tribe. However, and importantly, just as with all things in nature we also find some individual variation and the eggs produced by a particular garden warbler cuckoo are somewhat different from eggs laid by another garden warbler cuckoo. A similar variation is also found among host females. Obviously, a female cuckoo can increase the chances of her eggs being accepted by selecting not only the correct host species but also by selecting particular host clutches whose eggs most closely match her own (and avoiding host clutches with eggs that are too dissimilar). This is an intriguing possibility, but the evidence for such intra-host selection is poor and it is hard to imagine how such super-fine-tuned selection would work in the real world. We are still waiting for evidence that a laying female cuckoo ever looks at her newly laid egg in order to learn its appearance so as to later select host nests with the most similar looking eggs video recordings of reed warbler and common redstart nests have shown that females leave host nests without looking into them after laying (see chapter 5). Even if a female cuckoo had an innate 'idea' how her own eggs looked, the small number of host nests in her territory ready to accept an egg at the moment that she is ready to lay it would severely limit the scope of her choices. In order to test the exciting idea that female cuckoos select better matching host clutches one would need to radio-track individual female cuckoos, record which particular host nests she visited, and then confirm that the nest in which she eventually laid her egg indeed provided the best match available. Perhaps not surprisingly, this crucial test has yet to be made. Selection of particular host individuals within species may not only relate to the appearance of their eggs. Other clues which reflect their likely quality as foster parents may also be inspected by a cuckoo, for instance their body condition, or traits related to their attractiveness to partners such as the quality of their plumage colour.

No matter what mechanisms lie behind host selection, any particular female cuckoo always lays most of her eggs in the nests of a single host species. However, as is clear from the long list of species in whose nests cuckoo eggs have been found, it is not unusual for females to lay a few eggs in nests of the 'wrong' hosts. Indeed, molecular studies based on variation in DNA sequences have shown that specific tribes are not always a 'pure breed', but may rather have multiple, mixed origins. Reed warbler cuckoos in England and Denmark, for example, may have originated independently several times from other cuckoo tribes and this host switching is thought to be an important mechanism in the evolution of cuckoo tribes. For example, consider the case where a female cuckoo that usually lays her eggs in garden warbler nests faces a situation where these nests become scarce and hard to find, perhaps due to a sudden decline in the host population. The best she can do in such circumstances is to lay her eggs in the nest of alternative species in her territory, for instance a red-backed shrike. Or she lays her egg in a nest that resembles that of her normal host by mistake. If her eggs are accepted by the new host, and the resulting female chicks imprint on the new foster parents, they will return to parasitize red-backed shrikes next season – which could begin a new cuckoo tribe utilizing red-backed shrikes instead of garden warblers.

So far we have learned a good deal about the general principles governing cuckoo host selection behaviour, but how does this play out in real life? Which species generally make suitable hosts and which do not, and for what reasons? All of the species normally used are passerines (songbirds) which raise their chicks in a nest until fledging. Birds such as ducks, gulls or grebes are not suitable hosts since their young leave the nest soon after hatching and a cuckoo chick hatching in such a nest would be abandoned and quickly die.

And not all passerines are suitable hosts. Cuckoo chicks require an animal diet, so that species which feed their young exclusively on seeds, for example linnets and greenfinches, are avoided. In addition, the host species must not be too large. This is not because body size would be a problem in itself. Quite the opposite, because larger hosts would be, on average, better at bringing more food and defending the cuckoo chick against predators. The problem is that large hosts, like thrushes or crows, build large nests and have large eggs and chicks which a tiny cuckoo chick would be unable to evict from the nest. It would therefore have to compete with its host chicks for food, would fail to get enough care, and so would die.

Cuckoo in a sphere. A young cuckoo looking out of the globe-shaped nest typical for wrens.

Scientists have traditionally considered hole-nesting passerines as prime examples of unsuitable hosts. Yet, the real world is never black and white. On the one hand, the nest entrance holes are often small and this can have fatal results – a small egg could be easily smuggled inside (by projecting or 'shooting' the egg from the nest entrance, as finally confirmed by recent film footage) but once grown, the monstrous cuckoo fledgling may be easily trapped inside its prison cell and starve. On the other hand, a life spent in a hole is safer and more comfortable than in an open nest – the risks of predation are reduced, and the cavity works as a natural umbrella against the rain and cold. So, hole-nesting species which build their nests in cavities with large openings, such as common redstarts, are commonly used by cuckoos, while potential hosts breeding in holes with small openings, such as great tits and pied flycatchers, seem to be generally avoided.

Even potential hosts that build open nests and feed their young on insects may escape parasitism. Contrary to popular belief, cuckoos are not shy of people and may be found around human settlements. One hundred years ago it was not exceptional to find parasitized nests in villages and small towns, and there are many reports of cuckoos laying their eggs in nests built in the ivy covering houses bordering English village gardens. Olivaceous warblers breeding in Bulgarian villages and black redstarts breeding inside farm buildings in the Alps are also regularly used as cuckoo hosts. Modern towns and cities, however, seem to be avoided and several potential cuckoo hosts, such as blackcaps, robins and dunnocks, can escape parasitism when breeding in towns.

Other potential hosts, such as skylarks, escape cuckoos by breeding in open areas far from the trees used by cuckoos as 'watchtowers'. Nests that are well hidden may also be more difficult for cuckoos to find and are at lower risk of parasitism, although if the density of hosts is low female cuckoos may intensify their nest-searching behaviour so that even well-concealed host nests are targeted.

Several other potential host characteristics may prevent some hosts being used by cuckoos: their breeding range may not coincide with that of cuckoos (for example, meadow pipits breeding on Iceland where there are no cuckoos); their population size or density may be too low to support cuckoos; or they may breed before female cuckoos arrive from their wintering grounds (as is the case for the first clutches of robins or wagtails).

The males and females of brood parasites such as the great spotted cuckoo and the Asian koel may collaborate in smuggling eggs into host nests. The male may provoke the nest owners into chasing him away allowing the female to sneak her egg into the unguarded nest

unnoticed. This strategy works best for monogamous cuckoo species where a male and a female remain paired throughout the breeding season. However, our cuckoo does not form permanent pair bonds making such cooperation impractical. Males however may indirectly help females in detecting host nests. Males have been observed chasing each other over a reedbed and disturbing great reed warblers, which in taking flight inadvertently reveal the location of their nests to an observant female cuckoo hiding in her watchtower nearby. The male cuckoos' behaviour may make her job of finding host nests much less time consuming.

Searching for host nests. Female cuckoos spend many hours patiently watching potential hosts to discover the location of their nests so that they can lay their eggs.

Showing off. The pattern formed by the fanned tail feathers of a male cuckoo may function as a visual advertisement to females.

Arrival time. Cuckoos usually arrive in the second half of April as leaves begin to appear on the trees.

Food source. The growing leaves of oaks present a feast for innumerable caterpillars and other larvae, attracting many different birds including cuckoos.

Yummy. Hairy caterpillars often occur in abundant masses and are the favourite food for cuckoos. Most other birds avoid such hairy morsels.

In the air.
Territorial flights by males are common sights while cuckoos are mating.

•
Taking his perch.
A male cuckoo on the point of landing.

••
Combat wounds!
Conflicts between males are vigorous and can result in lost and damaged feathers.

•••
Dressed for success.
The elongated feathers on the lower back of cuckoos can be erected during mating and territorial conflicts, giving the impression of a strikingly large 'cocked-tail'.

A fight! Territorial conflicts between males can be intense and violent.

Fighting males.

Sunset. The silhouette of a male cuckoo during the mating period.

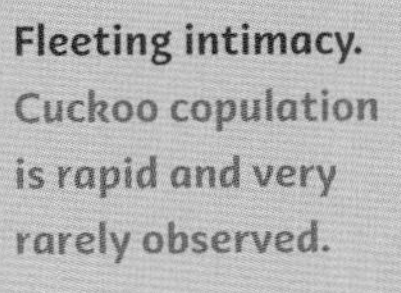

Fleeting intimacy. Cuckoo copulation is rapid and very rarely observed.

A tandem flight. Male and female cuckoos often interact during the mating period.

Cuckoo eggs

5. The stealthy deposit – Laying an egg into the host's 'nursery'

Once a watchful female cuckoo has detected a host nest, her job is far from over. Successful laying is a matter of perfect timing. The host female lays one egg each day until the clutch is complete. It is important that the cuckoo does not lay her egg too early, before the host has laid her first egg, because this will often result in the host deserting the nest. Furthermore, laying the egg too late, after the host has started incubating her full clutch, would also be disastrous for the cuckoo. The host's chicks would hatch before the cuckoo chick, outgrow it, and so gain a size advantage over the parasite chick making it much more difficult, or even impossible, for the young cuckoo to evict its foster 'siblings'. The female cuckoo must therefore time her egg laying perfectly so that the egg is laid in the four or five days during which the host completes her clutch. Because cuckoos can only lay one egg every other day, in contrast to the daily laying of small host species, it becomes even more difficult for her to fit into the host's narrow laying 'window'. Cuckoos have an extraordinary ability to time their egg laying within this short interval, but do sometimes make mistakes. Cuckoo eggs have been found laid into old nests, newly constructed nests without eggs, nests with incubated eggs and even in nests already containing host chicks.

Removing host eggs. A female cuckoo removes one or two host eggs just before laying her own, often devouring the host eggs in the process.

Bad timing. A small cuckoo chick and a freshly laid cuckoo egg in a great reed warbler nest. Not all cuckoo eggs are laid at the perfect time and this one was deposited into a nest which already contained a cuckoo chick, which will soon evict the egg – as it has already done to all of the host eggs.

When ready to lay, the female cuckoo glides out from her vantage point and lands at the selected nest. She often removes one or two host eggs, which are devoured. Why is the female cuckoo not only a parasite but also a predator? Perhaps if the female cuckoo did not remove a host egg, the host would notice the extra egg in her nest, alerting her to the uninvited guest about to destroy her home. But experiments have shown that hosts do not notice whether one of their own eggs has been removed or not after an experimental egg is added to the nest. Another reason may be that removing a host egg or two reduces the number of host eggs in the nest that the host can use to compare with the cuckoo egg, and so makes spotting the imposter egg more difficult. However, it is hard to see how this can fully explain egg removal by cuckoos – several host species can recognize alien eggs even when none of their own eggs are present for comparison. It may be that cuckoos remove some host eggs to facilitate incubation of their own egg. Most hosts are small passerines and a cuckoo egg is often a bit larger than their own. They would have difficulty covering too many eggs, making effective incubation harder and reducing the chances of them all hatching successfully. Furthermore, the fewer eggs that are in the nest, the less energy the hatchling cuckoo must expend to evict them. Eating some host eggs may also have a much more pragmatic benefit – eggs are enticing packages of calcium and energy, and female cuckoos needs lots of both to produce their own eggs. The thicker shells of cuckoos' eggs require more calcium to produce than do host eggs. Female cuckoos, however, should avoid removing too many host eggs because hosts are more liable to desert their nest if too many eggs are taken.

Too greedy. A female cuckoo has removed all of the host eggs before laying her own into this reed warbler nest. In such cases the host will always desert the nest and build a new one, so this cuckoo egg will never hatch.

Actual egg laying is often completed in seconds, before the cuckoo can be discovered by the hosts who would then become suspicious, investigate their nest contents more carefully, and become more likely to reject the foreign egg. Of course, this is the ideal and things may not go so well in nature – field observations have shown that cuckoos do not always follow the guidelines. They sometimes spend much longer at the nest, and may even chase away smaller hosts such as reed warblers in order to reach the nest. So, even though a particular strategy has been fine-tuned by natural selection, there will always be some variation in the behaviour of individual birds and actual circumstances at the time.

Caught in the act. A female cuckoo grasping a host egg is attacked by the great reed warbler which owns the nest before she is able to lay her own egg. Such attacks are sometimes sufficiently vigorous that the cuckoo is driven into the water and may drown.

Many hosts are very aggressive towards cuckoos and clearly regard them as a specific enemy, differentiating between cuckoos and other intruders near the nest, both dangerous ones, such as jays, and innocuous ones, such as doves. Some hosts also differentiate the grey and rufous female cuckoos. They often try to chase the cuckoo away from their nest to prevent her laying her egg. Smaller hosts, such as marsh warblers, often cannot successfully intimidate the larger intruder, but larger hosts such as great reed warblers actually pose a real threat. Most female cuckoos survive these attacks and only lose a feather or two in the battle. But experiments using dummy stuffed adult cuckoos, and direct observations of naturally occurring fights between cuckoos and hosts, show that hosts specifically target the most sensitive parts of the cuckoo's body, pecking directly at the eyes and nape. Drowned female cuckoos have been found in the water beneath great

Plumage mimicry I. Grey cuckoos (bottom) resemble sparrowhawks (top).

reed warbler nests and female cuckoos have been discovered brutally pecked to death under the nests of shrikes – telling evidence of hosts that have successfully prevented parasitism by those females forever.

It is this host aggression that has forced cuckoos into their secretive lifestyle and amazingly rapid egg-laying abilities. These adaptations have, in turn, probably led to hosts becoming increasingly attentive to their nests as a counter-adaptation. While this may seem like an evolutionary dead-end, cuckoos have fought back, but not by evolving new behavioural tactics. They have adopted a new 'outfit' that disguises them as a deadly enemy – the sparrowhawk. The similarity between this feared predator of small passerines (including cuckoo hosts) and the cuckoo itself is so close that an observer needs a good view and considerable experience to distinguish the original (the hawk) from its copy (the cuckoo).

The myth that cuckoos metamorphose into sparrow-hawks in the autumn survived in folk traditions till the 19th century. Both birds have a slim body, pointed wings, grey upperparts, barred underparts, and yellow eyes, bill and legs. Even their flight style is, from some viewing angles, confusingly similar. Because potential encounters between a cuckoo and a host near a nest are so brief, a host has barely a split second to make a detailed examination of the swift and secretive intruder and so runs the risk of mistaking a hawk for a cuckoo and dying as a consequence. Not surprisingly, some hosts such as reed warblers fear the cuckoo as if she was actually a hawk. However, some larger and less fearful hosts such as great reed warblers may take the battle to the next stage – these lionhearted birds respond to the hawk-like appearance of a cuckoo (specifically the yellow eyes and barred underparts) with increased aggression. Paradoxically, the parasite's disguise does not help the cuckoo in this case but rather the warblers – the 'fools' outwit the cheat.

Cuckoos are not common everywhere and hosts may not have many chances to learn about them from personal experience. Reed warblers follow the wise old dictum – it is always better to learn from the mistakes of others than from your own. These common cuckoo victims are vigilant observers of events in their neighbourhood – when a neighbours 'house' is visited by a cuckoo, reed warblers increase their responses to a cuckoo when she attempts to lay in their own nest. This is not a result of the general excitement caused

by any intruder – their own responses to a cuckoo are unaffected if they see their neighbours attacking a non-cuckoo intruder. How have cuckoos tried to escape the consequences of such social learning by their hosts?

Alerting the neighbours. A great reed warbler mobs a cuckoo, and so may inadvertently warn other warblers in neighbouring territories of the potential danger.

Cuckoos have brought another disguise to the battlefield to outsmart their hosts – in addition to the sparrowhawk-like females a new superficially kestrel-like form has evolved. This alternative morph is rufous and not at all like the hawk-like females. But even this tactic has not bamboozled the tiny brains of tiny cuckoo hosts – reed warblers now pay attention to identifying the cuckoo morph mobbed by their neighbours (for example, rufous) and afterwards attack the same morph (rufous) more intensely, while not changing their responses to the other morph (grey). This has interesting consequences – the morph that is locally more common is more often encountered, attacked and subsequently recognised by neighbours. An alternative, less common, morph has an advantage in being a 'rare enemy', is attacked less and so more easily sneaks its egg into a host nest.

But, success today can be the cause of tomorrow's failure – the increased success of the rarer morph leads directly to its increased occurrence in the population and to it becoming more often encountered, attacked and remembered by neighbours! Because hosts learn the most common current enemy it is always the rarer morph that wins. This explains why both grey and rufous morphs can coexist – their relative frequency see-saws back and forth leading to a natural balance between the two. Neither can evolutionarily 'erase' the other.

Host attacks may be an effective way to prevent parasitism in some highly aggressive hosts, but on the other hand it may also draw attention to the nest. The crucial thing is whose attention it is. The alarm calls of

Plumage mimicry II. Brown female cuckoos (bottom) resemble kestrels (top). However, recent experiments do not support the idea that brown cuckoo females mimic kestrels; instead, some females may simply keep the juvenile plumage till adulthood.

Communal mobbing. This spotted flycatcher mobs a male cuckoo, despite it not being a cuckoo host in this area of fish ponds where warblers are the usual hosts.

reed warblers attract neighbouring warblers who can help with nest defence and perhaps also confuse the intruder. A mobbing attack may, in theory, even attract a predator that would attack the cuckoo in turn and inadvertently help the poor hosts. But everything good may be bad when viewed from a different perspective – there is no reason to expect that any predator attracted to the fray would focus only on the cuckoo – a sparrowhawk may attack the hosts themselves, or a jay may take the confusion as an opportunity to eat their eggs. Blackcaps also 'call for help' when discovering a cuckoo near their nests and are often successful in attracting many birds of various species, but those are lazy and do not help at all. Overall, conspicuous mobbing may deter a cuckoo but also increase the likelihood of subsequent parasitism or egg predation by other female cuckoos.

Cuckoos quite commonly visit nests and eat host eggs and even chicks without actually parasitizing that nest. Some hosts then build a new nest in the vicinity and lay a new clutch, which again may be parasitized by the same female cuckoo. Hence, by acting as a nest predator, the cuckoo creates new opportunities for future parasitism. But, just as with many appealing theories regarding brood parasitism this idea that cuckoos can 'farm' nests remains just a hypothesis – nobody has so far followed individually marked cuckoos and hosts to confirm that it was the same female cuckoo which predated the first nest and then parasitized the second nest of the same host. Such a study would be logistically very demanding, but also very rewarding and may in fact reveal a much simpler world – one where cuckoos prey on host clutches and broods simply to feed themselves.

6. The great deception - Deceiving the host into incubating an alien egg

As we have seen, each female cuckoo belongs to a host-specific tribe, often producing eggs that match the particular host both in ground colour and spotting pattern. The production of eggs that mimic those of the host serves to avoid their being rejected by the host birds. Egg rejection is by far the most studied host adaptation against brood parasites. Once the hosts are convinced that they have been parasitized, they try to get rid of the foreign egg using one of several methods. First, the host can remove the egg either by pecking a hole in it and throwing it out of the nest or, if the host has a large enough beak, by just grasping it and throwing it out, without pecking it first. Second, they may simply desert the nest and build a new one nearby. Third, some hosts may build a new nest on top of the old one, burying both its own eggs and that of the cuckoo, in which case, the 'old' clutch cannot hatch and the host can continue laying a new one and thereby avoid the costs of parasitism.

Studies have shown that individual hosts rely on one of two cognitive mechanisms to recognise parasitic eggs. Some birds know the appearance of their own egg and will reject any egg that falls outside this template (which is either innate or learned) – an ability called 'true recognition'. Others will reject any egg that stands out from the rest of the clutch – so-called 'rejection by discordancy'. In several species, individuals may alternate between the use of both mechanisms.

What kind of cues do the hosts use to distinguish between their own and cuckoo eggs? Experiments with artificial eggs have shown that differences between own and foreign eggs in ground colour or spotting pattern, or a combination of these traits, may trigger host egg rejection. In some hosts, such as blackcaps, even subtle differences between host and parasitic eggs are sufficient for successful recognition of the foreign egg. Other hosts, such as common redstarts, require larger discrepancies to successfully recognise the alien.

A punctured egg. This cuckoo egg, laid in a reed warbler nest, has been punctured by the hosts – an efficient way of ridding themselves of the parasite egg.

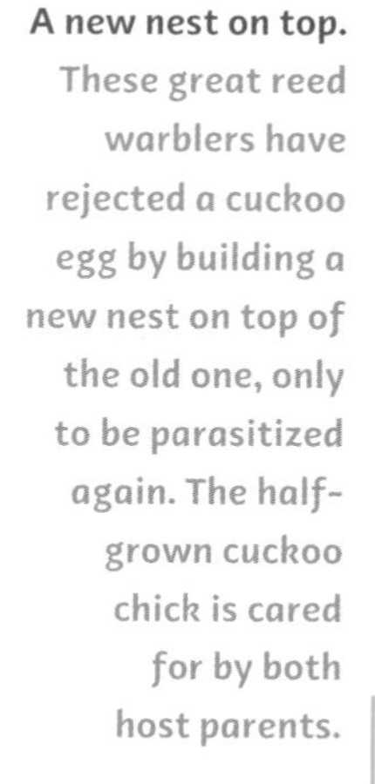

A new nest on top. These great reed warblers have rejected a cuckoo egg by building a new nest on top of the old one, only to be parasitized again. The half-grown cuckoo chick is cared for by both host parents.

Egg size mimicry. **Cuckoo eggs are much smaller than expected for birds of their size, an adaptation which makes it harder for hosts to detect the parasite egg. From left to right: eggs of a marsh warbler, cuckoo and turtle dove** (with a one cent Euro coin, for scale; Photo: Karl Schulze-Hagen).

In addition, cuckoo eggs are usually somewhat larger than the host eggs, even though they are unusually small in relation to the relatively large body size of the cuckoos that lay them. Based on body size alone, we would expect a cuckoo egg to be more than twice as big as it actually is. A female cuckoo is similar in size to a collared or turtle dove, yet parasitizes warblers which are 10 times smaller. Her egg is only one third of the size of a dove's egg and much closer in size to that of a warbler. This enormous reduction in egg size is an evolutionary masterpiece. Because hosts reject suspiciously large eggs and different hosts lay different sized eggs, we would expect to observe a match between the egg sizes of various hosts and their respective cuckoo tribe – and this is indeed the case. In some situations the mimicry in colour, spotting pattern and size is so exquisite (for example, in cuckoos parasitizing great reed warblers) that even experienced researchers can be fooled into thinking that a host nest contains nothing but host eggs, their error only becoming apparent when a nest visit after the eggs have hatched reveals the sole occupant to be a cuckoo chick.

Experiments have shown that certain areas of the egg's surface may be used as a key reference to separate own from foreign eggs. The blunt end is often the most visible part of an egg in the nest, and experimental eggs that differ from the host eggs in this region will be rejected more often than eggs differing (to the same extent) at the pointed end.

Commonly used hosts, such as reed warblers and meadow pipits, are more likely to reject cuckoo eggs when they spot a laying cuckoo at their nest. In these cases, hosts use two different cues in their decision to reject the parasitic egg – the presence of the egg and the presence of the cuckoo itself. Such a combination is referred to as 'stimulus summation' – meaning simply that 'two cues are more effective than one'. It appears that hosts examine the contents of their nest more carefully when they spot a cuckoo in the neighbourhood, which makes sense since if no cuckoos are seen in the vicinity parasitism is unlikely.

Some hosts may use other cues that a cuckoo has tricked them rather than the actual sight of a female cuckoo or suspicious egg. The wren is a small passerine that builds a neat bowl-like nest with a small entrance hole, made of grass, moss and small twigs. Wrens rarely eject eggs, and cuckoo eggs laid in wren nests never mimic the host. However, the older literature contains reports of wrens using an alternative defensive mechanism. When a female cuckoo inspects the nest of a wren she often inserts her head into the small entrance hole, thereby enlarging it a little before laying her egg. Wrens may spot this damage to the nest entrance and desert the nest as a result.

There is yet another host strategy which may be used to discover a foreign egg in the nest, and it relates to how orderly the homemaker is. Hosts often arrange their eggs in a specific way in the nest, maintaining this arrangement for many days. When a cuckoo (or a conspecific) parasitizes the nest, this arrangement is inevitably changed – the parasite removes one or more host eggs and jumbles the clutch when adding her own egg. The host can use the altered arrangement as an indirect cue that an uninvited guest has visited and act accordingly. In support of this idea, blackbirds and song thrushes often keep their eggs neatly arranged and also reject foreign eggs more often. However, experimentally shuffling the eggs in blackbird clutches did not affect their egg rejection decisions. It seems that altered egg arrangement alone is not a sufficient prompt for egg rejection – perhaps the relationship is the other way around and only females that are good rejecters are also good housekeepers.

The evolution of egg mimicry has resulted in an astonishing variation in egg colour and spotting pattern among the various cuckoo tribes. We find immaculate blue and white cuckoo eggs in the nests of common and black redstarts, respectively, each matching the host eggs to perfection. Even host species that lay eggs with variable spotting patterns do not escape parasitism. Bramblings, white wagtails, meadow pipits and several species of warblers and shrikes are all parasitized by cuckoos which lay highly mimetic eggs. However, in some other favourite hosts, such as dunnocks, the cuckoo eggs can be easily identified even by a colour-blind observer – densely spotted cuckoo eggs appear strikingly non-mimetic next to plain, pale blue dunnock eggs. Unlike the hosts mentioned previously, dunnocks accept all kinds of eggs placed in their nests, and without any selective pressure the cuckoo tribe parasitizing them has not needed to evolve mimetic eggs.

This mimicry vs. no mimicry dichotomy may sometimes be more confusing than enlightening. On closer examination, some cuckoo eggs that do not apparently mimic the eggs of their hosts may actually have evolved to lower the risk of host rejection after all. One such case is where the cuckoo egg is cryptic – it blends in with the colour of the nest structure or the nest environment. This could explain the very dark cuckoo eggs found in hosts breeding in cavities, semi-cavities or domed nests, such as wrens and robins. Crypsis may also evolve due to the selective removal of conspecific eggs laid by other cuckoos, as has been shown for one Australian cuckoo species.

Confusingly, hosts sometimes accept very dissimilar cuckoo eggs while ejecting ones that seem to be a perfect match to their own. This does not seem to make any sense to us – and this is exactly the point. Scientists believed for centuries that we humans live in a sensory world that is more similar to birds than to mammals – most mammals rely predominantly on their sense of smell, while birds and humans inhabit a visual world. We now know that the truth is rather different – the visual world of birds, the feathered dinosaurs, is completely different from our own and recent discoveries have shown that birds also have excellent olfactory abilities that match those of mammals! Unlike humans and other mammals, but akin to most other vertebrates and insects, many birds can perceive ultraviolet (UV) light, at much shorter wave-lengths (ca 320-400 nm) than we humans can register (400-700 nm). Indeed, the differences in UV reflectance alone between cuckoo and host eggs may be sufficient for hosts to recognise parasitic eggs, as has been demonstrated experimentally in blackcaps and reed warblers.

A non-mimetic egg. Hosts will usually reject cuckoo eggs that look different from their own. These great reed warblers, however, have accepted this pinkish, highly non-mimetic cuckoo egg.

While egg mimicry (and crypsis) is clearly a highly potent part of the cuckoo armoury, the parasite has other weapons as well. Cuckoo eggs have a remarkably thick shell compared to other species – a highly beneficial adaptation in many respects. Because cuckoos are considerably larger than many of their hosts, they cannot sit comfortably in the nest-cup while laying, but have to drop their more robust eggs from above. The strong eggshell also makes it more difficult for hosts to successfully peck a hole in the egg to remove it from the nest. Video recordings of olivaceous warblers, hosts of the cuckoo in Bulgaria, have actually shown that while some individuals recognize cuckoos eggs and peck at them, they lose their motivation when their efforts do not succeed in puncturing them. So some cuckoo eggs that are pecked without success are eventually accepted, incubated, and hatch successfully!

As expected, the higher the foreign egg rejection rate by a host is, the thicker the cuckoo eggshell 'shield' is for the respective cuckoo tribe. However, in many cases hosts are able to either grasp the cuckoo egg without puncturing it (larger hosts such as great reed warblers), or can successfully penetrate the cuckoo eggshell and remove the egg from their nest, showing that a thick eggshell alone is not always a sufficient adaptation to overcome host defences.

Another important cuckoo adaptation is the time required for their eggs to incubate. Recent studies have shown that cuckoos actually start incubating their eggs even before they are laid. This internal incubation within the female cuckoo's body gives the egg a head-start of 31 hours compared to a host egg. Cuckoos lay every other day but only take 24 hours to pass from ovulation and fertilization to the completion of a fully formed egg, ready to be laid. The extra seven hours of development arise from the higher temperature in the female cuckoo's body (ca 40 °C) acting as an internal incubator, compared to the external incubator of the host's brood patch (36 °C). As a result, the cuckoo egg will almost always hatch before the host eggs. This is important because of the relative ease and energy saving if the newly hatched cuckoo chick only needs to evict a motionless host egg rather than a struggling host chick.

High intra-clutch variation. One cuckoo egg (the palest) in a nest with three reed warbler eggs. In this case, the wide variety in pattern of the warbler eggs makes it much more difficult for the hosts to spot the differences in the parasitic cuckoo egg.

As we have seen, one of the best parasitic strategies that cuckoos have evolved is their mimetic eggs, and indeed the resemblance between cuckoo and host eggs can be astonishing. How can the hosts defend themselves against such exquisite trickery? In the same way that humans defend themselves against counterfeit

Low intra-clutch variation. One cuckoo egg (the palest) in a nest with three great reed warbler eggs. The similar appearance of the host eggs makes the cuckoo egg stand out much more clearly.

High inter-clutch variation. A selection of parasitized tree pipit clutches. This host is well-known for the huge variations in the colour and spotting pattern of the eggs of individual females, making it difficult for cuckoos to mimic their eggs. Egg collection of the Natural History Museum, Tring, UK (Photo: Bård Stokke).

banknotes. In order to recognize a mimetic egg, the host can standardize its egg production by laying eggs that are similar to each other as if they were 'printed' from an identical template so that the variation between eggs in a clutch is very low. In addition, they can produce intricate patterns and signatures that make it more difficult for cuckoos to mimic the eggs – especially if the 'inscriptions' are specific to each host female. In this way, hosts can detect a foreign egg more easily.

Some host species have evolved an additional line of defence in that each female lays eggs with markedly different colour patterns – in other words, they show high inter-clutch variation. One such species is the tree pipit, which as a species lays eggs with an extraordinary variety of colours and spotting patterns, but where each clutch laid by a particular female is remarkably uniform. Due to this high variation between clutches, the eggs of a particular cuckoo female may be good mimics in one host nest but poor mimics in another.

As previously mentioned, some host species accept every kind of cuckoo egg, mimetic or not, even though they are currently commonly used as hosts. The dunnock, which is frequently parasitized in England, the Netherlands and Belgium, is the prime example and it remains a mystery why it has not evolved any countermeasures against cuckoo parasitism. Perhaps it has only recently been adopted as a host and has not yet evolved any defences? This seems unlikely, because we know it has been a favourite host for centuries and Shakespeare mentions this in his writing. Perhaps it is because in many areas dunnocks produce several clutches each season and the first clutch avoids parasitism because it is completed before cuckoos arrive? The selective pressure for evolving defences could be reduced because the costs of parasitism for any individual dunnock are lower than if all their clutches were at risk. These are just some speculations among many, but such mysteries have intrigued cuckoo researchers for many years and spur us on to further research to test any new and intriguing ideas.

In summary, both cuckoos and hosts have evolved an arsenal of tricks and weapons to maximise the chances of successful incubation of their own eggs. If the cuckoo comes out on top of the situation, its egg will hatch in the host nest. But there is a still a long way to go before a cuckoo chick can successfully fledge from its adopted nest. Let us now look in detail at its life in the nest and examine the fascinating adaptations that it has evolved for the next stage in its life history.

• **As the season progresses.** Later in May the reeds have grown taller and reed warblers start to build their nests.

•• **Nest construction.** Great reed warblers build their nests in the reeds, using reed stems to provide a sturdy foundation.

••• **Here I am.** Male warblers, a great reed warbler in this case, sing frequently to advertise for females.

Sweeping over the reeds. A male cuckoo glides inconspicuously through the reeds.

••

••

•••

Unpopular. Cuckoos are chased away as soon as they approach the breeding territories of potential hosts, such as golden orioles (·), red-backed shrikes (··), lesser grey shrikes (···) and great reed warblers (pages 60–61).

• **On the alert.** Female cuckoos often perch motionless on a branch close to a reedbed, scanning for host nests.

• • **Sweeping over the reeds.** A female cuckoo flies swiftly over the reedbed seeking the whereabouts of potential hosts.

Lookout post. Even small bushes and tall herbs may be used as watchtowers.

Arrival. A female cuckoo inspects this great reed warbler nest before laying her egg.

What's in it for me? One or two host eggs are often taken by the female cuckoo before she lays her own egg. Sometimes females are caught in the act by hosts, like great reed warblers here.

Under attack.
Great reed warblers are aggressive birds and may put up a determined fight against female cuckoos.

A tasty nutritious package. Host eggs represent a rich source of calcium and other nutrients.

Even parasites may be parasitized. This egg-stealing cuckoo is being bitten by both a mosquito and a tick, both possible vectors of endoparasites such as malarial pathogens.

Greedy. The host egg is quickly devoured.

Rejected. This reed warbler has detected the cuckoo egg and punctured it. The next task is to carry it away from the nest without spilling its content on the host eggs.

Accepted. A great reed warbler incubates both her own eggs and a cuckoo egg.

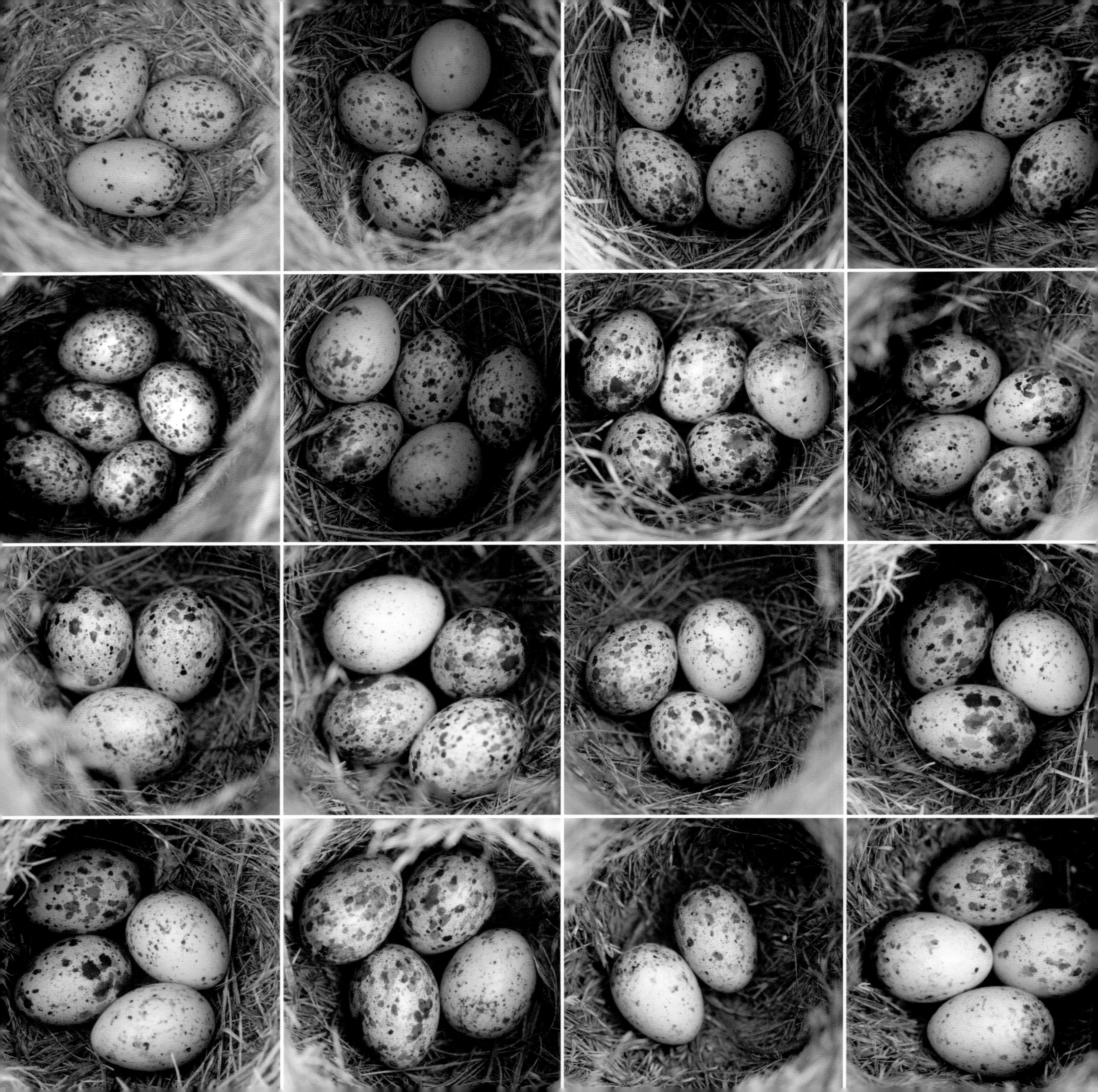

Left page. Clutches of great reed warblers containing cuckoo eggs.

Right page. Clutches of reed warblers containing cuckoo eggs.

Clutches of other cuckoo hosts each containing a cuckoo egg.

Top row from left: **Great reed warbler, marsh warbler, sedge warbler and black redstart.** Centre row from left: **Spotted flycatcher, blackcap, barred warbler and robin.** Bottom row from left: **Crested tit and common redstart (with two resp. one cuckoo egg).**

Cuckoo chicks

7. Killing them softly – Post-hatching behaviour

For 12 days the cuckoo egg has been a slumbering time bomb, and the emergence of the cuckoo chick triggers its detonation – with the familiar consequences that have been the stuff of literature ever since Aristotle described it 2300 years ago! The little parasite pushes his competitors – eggs and chicks alike – out of the nest, letting them fall, one after another, to their deaths, until it becomes the sole occupant of the host nest. Well into the 20th century many ornithologists refused to believe that a naked, blind, helpless chick – a creature that cannot even maintain its own body temperature and can barely raise its head – could perform such a feat of strength.

Hard load. This cuckoo chick, now two days old, is pushing a freshly hatched common redstart chick over the edge of its own nest.

Not long before, it had been an embryo and now it can wrestle and defeat the contents of its host's nest. It can even push host chicks over the edge of the nest that are heavier than itself. Imagine lifting a 100 kilogram bag of cement on your back over the edge of a hole three metres deep, without any sort of previous training. We wonder at the muscular power of a humpback whale breaching the surface of the sea, or of a peregrine falcon in its 300 kph diving stoop, but like for like the sustained effort of this fragile, pampered baby parasite is much more impressive.

Eviction completed. The cuckoo chick evicts host eggs and chicks using its concave back, rugged skin, and wing tips full of sensory nerves. The reed warbler egg is heaved into the abyss.

As early as 230 years ago the world-famous discoverer of the smallpox vaccine, the English doctor Edward Jenner, followed shortly after by his French medical colleague Antoine Lottinger, carefully observed and described a cuckoo chick's eviction of its host eggs and young. Jenner's study, published in 1788 in the *Philosophical Transactions of the Royal Society of London*, is a pioneering landmark in the precise observation of animal behaviour. He wrote, 'It is wonderful to see the extraordinary exertions of the young cuckoo, when it is two or three days old, if a bird be put into the nest with it that is too weighty for it to lift out. In this state it seems ever restless and uneasy'. Even one hundred years later many scientists still rejected this description as 'Jenner's preposterous account of the young cuckoo'. Many early zoologists thought Jenner was simply a liar, and that his precise observations and records must have been inventions. Yet all they had to do to discover the truth was to stand near a host nest themselves and watch patiently – just as Jenner did. Just over 100 years ago the first photographs of the event clinched the matter, at once silencing the sceptics.

But first let us look at the actual sequence of events in more detail, some aspects of which are very familiar while others have only recently been discovered. The shell of a cuckoo's egg is especially strong and requires twice as much work and strength to break open as the

It couldn't care less. Cuckoo chicks continue to evict host eggs whether the reed warbler foster parents are present or not. This young cuckoo ignores any food offered by its foster parents while it evicts their eggs – its only aim at this stage being to get rid of its competitors.

egg of a host. For example, the average cuckoo chick in a great reed warbler nest takes seven hours to hatch and needs to deliver around 250 mighty blows of its bill to chip and break the egg shell. The hatchling needs a particularly strong egg tooth and extra powerful neck muscles to complete the task and the chick must be completely exhausted by the end.

After a rest of just 24–36 hours the blind cuckoo chick pushes itself under a host egg or chick so that it comes to rest in a special, concave area on its back. This process is aided by its tiny wings, while its legs are braced against the nest wall and its head is bent downwards. In order to properly 'feel' its progress, the skin of the blind little killer is packed full of sensory nerve cells, particularly on its back and wings, as Jenner had earlier reasoned in a brilliant piece of deduction: 'With these (the extremities of its wings) I have often seen it examine, as it were, an egg and nestling before it began its operations; and the nice sensibility which these parts appeared to possess seemed sufficiently to compensate the want of sight, which as yet it was destitute of'. The little labourer often succeeds in heaving its freight onto the nest rim after only several attempts. The chick is prevented from falling back down into the nest by its widely spaced legs and by its tight grip on the nest material using its claws. With jerking movements of its whole body it finally tips its burden over the edge of the nest – its whole body commonly trembling during the exertions of this heavy work.

It takes two minutes for the average cuckoo chick to evict a single common redstart egg from the host nest. Even so, it takes several hours, or even days, for the parasite to eliminate all its rivals. This is because many eviction attempts are unsuccessful and the egg in question may fall back into the nest cup so that the cuckoo has to start again from the beginning. The chick evicts most eggs in the afternoon than in the morning, working away whether the foster parent is brooding it or not, while the hosts look on with indifference and make no attempt to save their progeny. In extreme cases, the cuckoo chick may still be trying to evict eggs that have fallen back into the nest some seven days after it hatched. At last it may give up and cuckoos that have not managed to heave all of their step-siblings over the side are then forced to grow up alongside them. While this situation is not particularly common it is more often seen in cavity-nesting hosts, such as common redstarts.

Evicting host eggs and chicks is demanding and risky. The work involves increased levels of oxidative stress, and eviction efforts in redstart nests have been shown to reduce the growth rate of cuckoo chicks by 20%. Cuckoo chicks are often so focused on their job, or are so exhausted by the work involved, that they ignore offers of food from their foster parents, and their food intake may fall by up to 30%. However, after completing its brutal task the cuckoo nestling quickly eats enough to compensate for its retarded growth, reaching the same weight at fledging as chicks that were helped by researchers removing all of the host eggs, so sparing them the costs of eviction. Nevertheless, the high costs of evicting the host eggs and chicks from common redstart nests means that cuckoo chicks fledge about one day later compared with those helped by an experimenter doing this job for them. Although the

Unsuccessful evictions. Sometimes a cuckoo chick is unable to evict (some of) its step-siblings and is forced to grow up alongside them, in these cases in a red-backed shrike (·), great reed warbler (··) and a common redstart (···) nest. Such events are not rare in the cavity nests of redstarts because of the steep inner walls of the nest cup; the evicted host chicks sometimes even climb back into the nest cup.

growth cost of eviction is recoverable, there are costs that cannot be recouped – a chick may die when it inadvertently falls out of the nest while struggling with an awkward host egg or chick, as sometimes happens in the open-nests of hosts such as reed warblers.

These costs of eviction prompt us to wonder why eviction behaviour exists at all? An obvious answer is that by killing its competitors the cuckoo can benefit from food brought to the nest that would otherwise be eaten by its competing host siblings. But this assumption may be incorrect, and there are dozens of well-documented cases where hosts have successfully fed both their own chicks and a cuckoo to fledging. In general, a larger brood does not necessarily mean less food per chick. In fact, chicks of the brown-headed cowbird (an American brood parasite) actually receive more food when sharing the nest with the host brood. When there are more chicks in the nest, the parents work harder to gather food – up to their foraging limit, of course – and more food is delivered per chick. However, the cowbird is better at competing for this food than are the host chicks and so gets more than its fair share.

According to the animal behaviour textbooks, cuckoo chicks are an example of a 'supernormal stimulus' whereby a particular stimulus elicits a stronger response than a standard one. In the case of the cuckoo, the imposter is so 'attractive' that foster parents bring more food to it than to a similarly-sized chick of their own. There are even observations of birds other than the foster parents feeding a cuckoo fledgling hatched in another bird's nest, begging the question as to why a cuckoo chick always goes to the effort of evicting the host progeny at all. Perhaps the extra food that would be brought to a larger brood, not 'pruned' by an evicting cuckoo chick, may be offset if the cuckoo becomes infected by diseases of the host chicks, or infested by their ecto- and endoparasites. These possibilities are as yet untested, but we do know of other risks associated with sharing a nest with the host brood. The host chicks fledge at a smaller size and so would always fledge earlier than the cuckoo, so the host adults would then only pay attention to the fledged chicks outside the nest, effectively deserting the cuckoo which would then die unattended in the nest.

As we have seen, the costs to the cuckoo chick of evicting eggs and chicks arise through reduced feeding rates and slower growth in the first days of life and delay its fledging by a whole day. But later on in its long nestling period, some 19–24 days in total, much higher costs can arise. Cuckoos experimentally raised alone always fledged in very good physical condition, whereas only half of those sharing the nest with common redstart young survived to fledging and – even worse – were in very poor condition. We know that low fledging weight in birds eventually leads to increased post-fledging mortality, another reason why evicting the host progeny is more beneficial than not evicting them, and another driver for the evolution of eviction behaviour.

Wouldn't this effort be reduced if the task of eviction were taken over by the chick's biological mother, the female cuckoo? We know that female cuckoos eat both host eggs (when laying their own egg) and chicks (in nests too advanced for parasitic egg-laying). If a female cuckoo consumed all the host eggs during her laying visit she would ensure that her own chick need not spend many hours or days in costly eviction attempts, doing what the female could do in a few seconds. From this perspective, eviction by the cuckoo chick makes no adaptive sense at all. But the 'female kills' strategy cannot win over the 'chick kills' strategy because cuckoo hosts always desert their nest if the clutch size is reduced to a single egg. In contrast, after the eggs have hatched the hosts will accept any situation, be it only one nestling or several, and the alternative 'chick kills' strategy remains effective. In addition, interactions between a cuckoo chick and older, larger and stronger host chicks may prove fatal meaning that the cuckoo must evict them as soon as possible after hatching, before they become a potential threat. The hosts' behaviour therefore both constrains and facilitates the persistence and timing of cuckoo eviction behaviour.

The success of eviction attempts depends strongly on nest cup architecture. Surprisingly, it does not matter how deep the nest cup is – it is only the steepness of the nest cup walls that helps or hinders eviction. Even a deep nest can effectively be quite shallow if it is wide with gently sloping walls, making eviction relatively easy. Conversely, a narrow, shallow nest can have nest walls sufficiently steep to prevent a cuckoo chick from easily pushing the host eggs or chicks out of it.

8. Rearing an ever-hungry monster – Life as a nestling

From a tiny helpless hatchling to a gargantuan fledgling, the cuckoo chick undergoes remarkable and speedy changes in its colour, size and behaviour. The newly hatched chick has pink skin and an orange gape but within a couple of days the skin turns blackish and the gape darkens to orange-red. What causes these conspicuous changes and why do they occur? The bright red gape was formerly thought to be irresistibly attractive to the host foster parents, but in fact appears to have no effect on the amount of food that they bring to the cuckoo chick. Having been startled by many cuckoo chicks in the wild, we wonder if the red gape, hissing sounds, and aggressive behaviour might together act as an anti-predator strategy rather than as supernormal stimuli for chick provisioning?

The growing monster. A young cuckoo about a week old. Soon it will grow to fill the whole nest.

Changing from pink to blackish. The skin colour of young cuckoos changes during the first few days of their lives.

The feather sheaths start to grow four days after hatching and erupt at seven days of age, leaving a week-old cuckoo chick looking like a small hedgehog. The chick is fully covered with feathers by the time it is nine days old and its trademark one or two white spots on the top of the head appear soon after. Except for these spots and the white tips to the tail feathers, the rest of the plumage is uniformly grey to rusty-brown.

Like the host progeny, the cuckoo chick also grows quickly. Contrary to expectations, cuckoo growth does not depend on the host's body size and nestling cuckoos put on weight at a similar rate whether in the nests of small hosts such as wrens or leaf warblers, or in the nests of larger hosts such as great reed warblers or shrikes.

The behaviour of the chick also changes dramatically as it grows. For the first day or two, it remains generally passive and even raising its head to beg seems to exhaust it. However, this weakling soon transforms into a frenetic character. About a week after hatching it mutates from 'Mr Helpless' to 'Mr Dangerous'. When an intruder approaches the nest the cuckoo chick bursts into a remarkable performance – it stands erect, opens its colourful gape wide, and moves its head repeatedly back and forth. If handled, it pecks vigorously at everything around it and – as researchers discover to their surprise – produces repulsive smelling liquid faeces that leave a difficult, stinking stain. For the unfortunate researcher it is easy to imagine that the cuckoo's performance may well be an effective weapon against a variety of predators. An experimental study has shown that the liquid faeces repelled mammals such as dogs and cats, as well as birds including raptors and

Stay away from me! The young cuckoos adopt a threatening posture if disturbed at the nest in order to intimidate predators.

owls, while scavenging birds, such as magpies or crows, were indifferent to the smell. What are the chemical constituents responsible for this repulsive effect? Do cuckoos use a similar chemistry to hoopoes, which are famous for their ability to produce a stinking secretion? Nobody knows as yet.

The growing chick also starts to pay more attention to its hygiene and after 11 days starts to preen itself

Keeping clean. During its long stay in the host nest, the cuckoo chick devotes a great deal of time to preening its feathers.

and stretch its wings. Another behavioural adaptation relating to self-defence emerges at 16 days of age – when danger approaches the chick produces a rattling call similar to the alarm call of a hawk. Meanwhile, its begging behaviour develops rapidly. At first, the chick just raises its head, with much trembling, and opens its gape, making only weak begging calls. However, as the days go by the begging calls become louder and louder, until the incessant calls of older nestlings can be heard from dozens of metres away such that one wonders why all cuckoo chicks are not eaten by eavesdropping herons or harriers!

Remarkably, the structure of the begging calls is specific to different cuckoo tribes and, for example, reed warbler cuckoos make a different sound to dunnock cuckoos. Interestingly, if we move a cuckoo hatched in a reed warbler nest into a dunnock nest it will then beg like a dunnock chick. How is that possible when there are no host chicks in the nest to learn from? At first, the young cuckoo probably performs a variety of different begging calls and then repeats only those that get the most response from its foster parents.

This rapid, fine-tuned learning of the 'correct' begging call structure following reinforcement, contrasts with their fixed responses to host alarm calls. When they spot a polecat or jay near their nest, avian parents alert their offspring by giving alarm calls and the chicks respond by falling silent and crouching in the nest cup. Reed warbler cuckoos are behaviourally pre-programmed to react only to reed warbler alarm calls. They not only fall silent and crouch, but also open their orange-red gape, perhaps to startle the predator. When raised by other host species in cross-fostering experiments, cuckoo chicks did not respond to the alarms of their new 'parents', but instead remain rigidly responsive only to reed warbler calls. Chicks of other hosts, such as robins and dunnocks, are similarly inflexible. It makes sense that responses to alarm calls are 'hard-wired' because predation is a 'one-time' event with terminal consequences. There is no second chance to allow chicks to learn the correct response and innate pre-programming to species-specific alarm call structure is the only effective protection. Curiously, common redstart cuckoos do not behave in this way and will respond to the alarm call of any host species.

Always hungry. Four, or even five, open gapes in an un-parasitized reed warbler nest provide a greater visual stimulus than the single cuckoo gape in a parasitized nest.

Cuckoo chicks beg even when the hosts are away from the nest, with clamouring, distinctive 'si' calls repeated every 0.5–5 seconds. The hungrier the chick, the more intense is this 'host-absent vocalization'. However, experimental playbacks of recorded host-absent vocalizations did not increase the amount of food brought by reed warbler foster parents. It appears that the call may not be a begging call after all, but may actually serve to establish a bond between the cuckoo chick and its foster parents, a long-lasting bond that will have to persist for up to about three weeks after fledging.

Birds feeding their chicks need to know how hungry the brood is and then collect and deliver the required amount of food. To do this effectively parents must integrate both visual and vocal information. First, the total gape area of the brood – the sum of the gape sizes of all of the chicks in a brood – increases both with brood size (the number of chicks) and with chick size (the age of the chicks). Older, larger chicks need more food and gape area therefore reflects their degree of hunger. But this hunger signal is not sufficient information for the parents – two broods with the same number of chicks of the same age can differ in how hungry they are. Therefore both parents and chicks need an additional signal and this is supplied by the chicks' begging call rate. So, while visual signals can provide a rough measure of the brood's total food requirement, vocal signals provide a fine-tuned measure of how hungry the chicks are at any moment in time.

Why do cuckoos beg at such a super-fast rate, sounding like a whole brood of host chicks? Passerine nestlings have huge gapes relative to their body sizes. In contrast, the cuckoo gape, although larger in absolute terms, is quite small relative to the parasite's large body size – a body that soon outgrows a whole brood of three or four host young to finally weigh as much as two fledged host broods. The cuckoo chick compensates for its insufficient, subnormal visual stimulus with a supernormal, exaggerated vocal stimulus to persuade the hosts to satisfy its insatiable appetite.

Twist and shake. This fully grown cuckoo vigorously shakes one wing in order to increase the foster parent's feeding rate, a white wagtail in this case.

In general gape area and call rates stop increasing almost a week before fledging, while chick weight continues to increase, requiring increased food delivery. Clearly, another begging stimulus must be added to the cuckoo's arsenal and two-week-old cuckoo chicks do in fact add another ruse to prompt increased feeding rates – wing-shaking. The chick raises one wing, and shakes it vigorously, as if waving 'Hello!' Unlike the majority of birds that shake both wings during begging, cuckoos always shake only one wing, the one facing an approaching fosterer. Wing shaking is presumably energetically demanding and therefore may tell the hosts 'I am a healthy, vigorous, high-quality chick worthy of more food!' Moreover, the cuckoo underwing has a conspicuous patch of white feathers, perhaps to make the signal more conspicuous.

Fostering a cuckoo is costly to the hosts and the necessary increase in feeding rate takes both time and energy. Shortly after hatching, the cuckoo chick requires more food than a single reed warbler chick of the same weight and by the time it is one week old, it needs more food than a whole warbler brood at fledging, when provisioning reaches its peak. Across the whole nestling period, from hatching until fledging, the cuckoo consumes at least twice as much as an entire host brood. The diet of the growing cuckoo consists mostly of flies, spiders, aphids and other invertebrates – a similar menu to that of host chicks, except for one striking difference. Cuckoos receive many more aphids, not a very economical prey to collect. Why would the hosts go to trouble of collecting such large numbers of such small creatures?

It seems that the frantically-working hosts have been moulded by natural selection to gather the maximum amount of food that would be needed by their own chicks – but not more. Once the cuckoo outgrows the average-sized host brood, three or four young at fledging, the average prey size decreases because of the increasing proportion of small prey, especially aphids and small spiders. The foster parents become overworked and are forced to reduce both their foraging selectivity, and perhaps also their future survival or reproductive chances as they tire in their struggle to feed the voracious parasite chick.

Every cuckoo chick appears to have its own personality. Some produce host-absent vocalisations almost continuously, while others never give this call at all. Some chicks start to use wing-shake begging at an early age, almost a week before fledging, while others only start just before leaving the host nest. Are these striking differences related to the sex of the chick, or its body condition, or to how well the host pair can feed them? As yet, nobody knows and another avenue of research opens for the future.

In need of attention. Some cuckoo chicks have a distinctive white spot on the forehead, which – in combination with their orange gape – may help to stimulate the foster parents to bring food.

What's on the menu? Great reed warblers feed chicks with a variety of food items, including small fish and frogs.

9. The attractive alien – Fledging and post-fledging behaviour

Unless the cuckoo chick is devoured by a marten, heron or snake, or dies due to poor weather, it will fledge safely provided that no misfortune befalls its foster parents. The young cuckoo needs around 18–24 days to grow fully and finally leave the nest, twice as long as the chicks of most host species, and weighs between 70 and 110 grams, depending on the particular host species. Remarkably, in contrast to the hundreds of studies focusing on the cuckoo's egg stage, only a single study (by Ian Wyllie) has investigated events during the post-fledging period. Cuckoos fledged from reed warbler nests are attended by their foster parents for another two or three weeks after having jumped out of the host nest. During the first couple of days the cuckoo cannot fly and just flutters and clambers about in nearby bushes and trees – begging for food from any birds it meets, even species other than its foster parents. After mastering the ability to fly, the fledgling travels several hundred metres, always followed by its foster parents. This is further than the distances moved by fledgling warblers, suggesting that post-fledging care may pose another substantial cost to the hosts. Sometimes reed warblers even follow their cuckoo fledgling deep into a nearby forest to feed them – an environment in which nobody would expect to meet these normally marsh-dwelling birds!

Friend or foe? A cuckoo chick is cared for throughout its time in the nest. As soon as it leaves and ventures into the reedbed, the fledgling may be regarded as an enemy and be attacked by its reed warbler foster carers. In this case the fledgling ceased attacks from its fosterers by returning to the security of the nest.

The big baby. The cuckoo chick soon greatly outgrows its foster parents, a reed warbler in this instance.

The fledged cuckoo is exposed to many and various dangers in addition to hungry predators seeking a succulent meal. Chicks that fledge in reedbeds must be able to swim, for they may have to leave the nest in a hurry and suddenly find themselves falling into the water – and indeed they can swim astonishingly well! One young cuckoo, only just able to fly, was seen

making an emergency landing almost in the middle of a Finnish lake, but swam for several minutes without any signs of distress, reaching the shore quite comfortably.

While the young cuckoo is still in the nest, it is invariably fed by its host parents. However, as soon as it leaves something strange can often occur – instead of feeding it, the foster parents may start to mob the cuckoo and attack it aggressively – so long as it makes no sound! As soon as it starts to beg again the hosts switch from attacking it to feeding it! The foster parents appear to be in the grip of conflicting tendencies to either feed an offspring or to attack a parasitic enemy. The attractive alien seems to deploy its begging calls as a 'remote control' to turn on the parental responses of its foster parents.

10. Escaping the workload – Recognising the parasite chick

Given the very obvious differences between a cuckoo chick and the host chicks, we would expect that hosts would easily recognise the imposter and give up on their misdirected efforts. Instead, we see perhaps the most mind-boggling enigma of the whole cuckoo story. While many hosts are able to detect even highly mimetic cuckoo eggs, they are totally blind to the extraordinary differences between their own and alien chicks. Cuckoo nestlings have different coloured skin, gape and feathers; they beg differently, behave differently, smell differently; and are much bigger – several times larger than even the foster parents themselves. Yet the foster parents pay no attention to any of these many contrasts that seem so striking to us humans. When chicks of various other species were placed in the nests of a variety of cuckoo hosts, no chick discrimination was ever found, no matter how different the imported chicks were.

Why have the hosts of our cuckoo evolved the ability to discriminate and reject alien eggs but are still unable to do likewise with alien chicks? It appears that the benefits of rejecting parasitic chicks are always lower than of rejecting their eggs. The mere fact that eggs come first and chicks come later inevitably means that the rejection of chicks is preceded by the investment already wasted by the host during incubation. Therefore, the evolution of a chick rejection strategy requires stronger selection pressure than the evolution of the ability to reject an egg. Moreover, the rejection of the egg directly decreases the effective parasitism rate at the nestling stage. Imagine a host that rejects all foreign eggs – no matter what the egg parasitism rate is, the effective chick parasitism rate would be zero, so no selection for anti-chick strategies would occur. In other words, a host that can always detect and reject parasite eggs has no need to evolve the ability to recognise parasite chicks.

Generally, cuckoo eggs are 'rare enemies' simply because only a small proportion of nests are attacked by cuckoos. Parasite chicks are even rarer enemies, because many cuckoo eggs never hatch due to host rejection or predation. This means that egg rejection behaviour will most likely evolve and spread before, and faster, than chick rejection behaviour – and so acts against the evolution of the latter! In consequence, the major factor responsible for the rarity of chick discrimination might actually be a behaviour of the host itself, namely its anti-egg adaptations. This paradoxical, yet simple, 'rarer enemy effect' suggests that across various host-parasite systems we should find chick discrimination only in hosts that accept all foreign eggs – and indeed, all of the hosts that are known to reject parasite chicks by desertion or ejection generally accept every natural parasitic egg.

However, a few studies have discovered exceptions to this rule. For example, a study in southern Moravia, Czech Republic, found some 15–20% of cuckoo chicks dead in the nests of reed warblers. All the dead chicks were about 14 days old, and had apparently been deserted by their foster parents. In some cases these hosts even disentangled the original nest and used the recycled material to build a new nest nearby while the cuckoo nestling still begged for food in vain! But, this clear rejection of cuckoo chicks seems to be unique among European cuckoo hosts.

What drove the warblers' decision to desert the alien chicks in this case? We know from experiments that warblers feed any chicks placed in their nests, so nestling discrimination could not be caused by the many differences in appearance and behaviour of the young cuckoo. Reed warbler chicks leave their nest at 11 days of age when the young cuckoo is still poorly developed – needing 18 days or more to fledge, nearly twice as long.

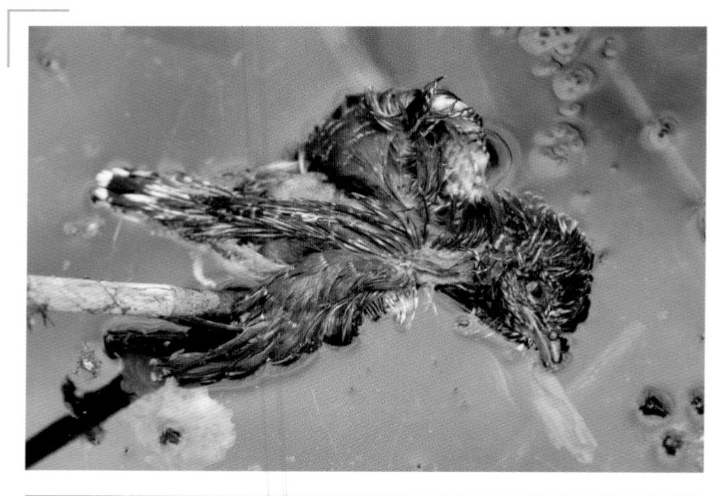

Unlucky weather. During periods of heavy rains, chicks may die due to hypothermia or even fall out of the nest and drown.

Left to die. Sometimes a cuckoo chick may be abandoned by its foster parents (here reed warblers), especially after the usual nestling period of the host chicks has expired.

Conceivably, this time delay collides with a reed warbler's 'internal clock' that programs the host's 'expectation' of when fledging should occur, leading to some reed warblers being able to 'discriminate without recognition'. They give up on the cuckoo because it remains far too long in the nest despite all their efforts to coax it to leave. In this case, how can young cuckoos survive at all in such a reed warbler population? Variation provides an answer – just as some host individuals reject eggs whereas others accept them, some hosts are more sensitive than others to the conflict between their 'expected' and experienced duration of nestling care.

We do not even know if this form of chick discrimination is a specific anti-cuckoo defence or a more general response to tardy chicks. Perhaps restriction of the period that warblers are willing to commit to nestling care evolved due to a parent-offspring conflict over fledging time decisions. Chicks may 'want' to stay in the cosy nest and be fed for longer than their parents are 'willing' to spend. At any rate, the effect is that some host pairs are spared making a prolonged investment in a parasitic chick.

It's a hard life. Recently fledged cuckoos may be taken by various avian and mammalian predators.

•
By late June, the reeds have grown much taller.

••
Hatching. This tiny cuckoo chick has just hatched in a reed warbler nest and its egg tooth is easily visible.

The moment of eviction. With incredible force and endurance, this seemingly helpless cuckoo chick expels the host eggs one by one (and may even evict another cuckoo egg or two host eggs at the same time on rare occasions).

pages 96–103

A growing monster. The cuckoo chick monopolizes the care of its foster parents and grows quickly. It leaves the nest before it is three weeks old, leaving the once tidy nest a mere wrecked platform.

Demanding. A big dragonfly is brought in by the great reed warbler foster parent for the ever hungry cuckoo fledgling.

July. The reeds have reached their maximum height.

pages 107–108

In the water. Young cuckoos just out the nest are poor fliers, and some may fall into the water below. This one manages to swim ashore and is fed by its foster parents as a reward.

Still demanding. Post-fledging, young cuckoos are still cared for by their ever eager foster parents.

Waiting impatiently. A young cuckoo perches in the reeds while the foster parents are away searching for food.

pages 110–113
Feeding the fledgling. To our eyes, a small reed warbler feeding a huge cuckoo chick remains both a bizarre and comical sight.

Reedwarblers sometimes follow their foster childs even into the woods.

Flying into the future. A newly independent juvenile cuckoo flies into the next stage of its life.

The complex co-evolution of brood parasitism – theories and research

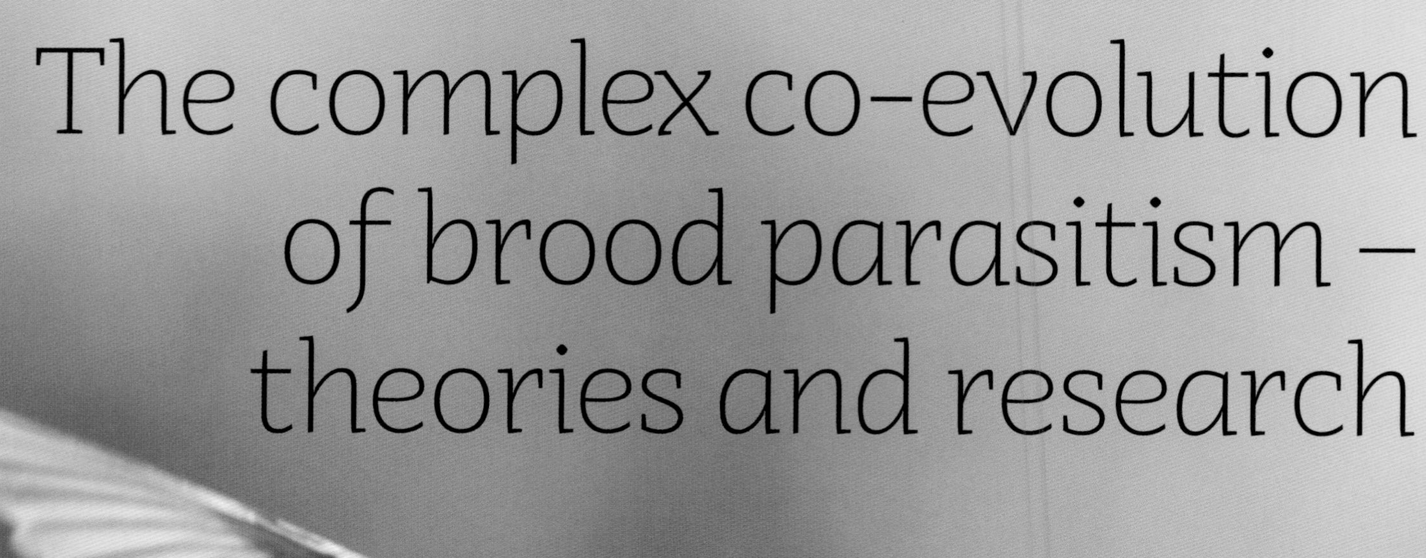

11. The battle of evermore – Arms races in space and time

The conflict of interest between cuckoos and their hosts escalates into a true arms race, driven by costs and benefits in which traits such as egg rejection and egg mimicry improve by evolution through natural selection. The basic principle of the famous arms race metaphor arose in the post WWII cold war period and serves as a good analogy for the cuckoo-host 'war'. When hosts (for example, the USA) developed a more effective egg recognition ability (say, a missile), cuckoos (for example, Russia) needed to counter this by developing more effective mimicry (say, a missile-detection system). Each improvement in one party's arsenal inevitably 'forced' the other to invest more into measures that would nullify this progress. And so on, and on.

Evolution by natural selection rests on a genetic basis. Traits like hair colour, height, and skin colour in humans, or milk yield and coat colour in cattle, all show variation as a result of both genes and the environment. The assembly of traits that we can see is called a phenotype. Hence, cows with brown coats and long horns represent one phenotype and those with black coats and short horns represent another. The underlying genetic basis for any set of traits is called the genotype. For natural selection to work and evolution to take place, there must first be variations in phenotype that are heritable and can therefore be transferred from one generation to the next. Gene copies can only be transferred in this way, and variation in phenotypes must be directly associated with heritable genetic variation. Second, and vitally, there must be differences in reproductive success between different phenotypes.

Egg mimicry in cuckoos evolves through natural selection because non-mimetic eggs are rejected by the hosts and die. Those female cuckoos laying the best mimics are more successful, leave more offspring, and pass more of their gene copies to the next generation at the expense of cuckoos producing poor mimics. Good mimics become more common in the next generation and good egg mimicry increases generation by generation, especially if new genetic variants producing increasingly good egg mimics arise by further mutations.

We have seen how parasites and hosts have evolved various means of attack and counter-measures, respectively. The first host line of defence is aggression towards the adult cuckoos approaching host nests – obviously, the best protection against parasitism is not to allow the nest to be parasitized in the first place. Cuckoos then fought back – they scare their hosts by mimicking deadly raptors, behave secretively, and lay their eggs super-quickly. Using these strategies, they can often overcome the first line of host defence. Yet, the host can still win by discriminating the cuckoo egg and ejecting it, building over the nest, or deserting the nest altogether. But many cuckoos overcame even these second line defences. They lay eggs that perfectly match host egg size, shape, background colour and spotting pattern. This often works and the cuckoo egg is accepted and the chick hatches. At this point we would expect to see a third line of host defence evolve. However, as we have seen, hardly any European cuckoo hosts have evolved the ability to recognise cuckoo chicks. This surprising absence of chick discrimination explains why cuckoo chicks bear no resemblance to host chicks at all. A potential fourth line of host defence, discrimination against fledglings, is theoretically feasible, but has not yet been studied in cuckoos.

Generally, all good things also have a bad side – any single adaptation, both of cuckoos or hosts, is likely to come at a cost. For example, strong cuckoo eggshells may provide an effective shield against host attack, but an eggshell that is strong on the outside is also strong on the inside. Therefore, the cuckoo hatchling spends much more time and energy to break through its eggshell than a host chick does. Egg rejection by hosts is a potent weapon against cuckoo eggs, but recognising a cuckoo egg requires host individuals to develop and maintain the necessary neural circuits, which also costs time and energy. Furthermore, an egg rejecter may make errors – she can mistakenly eject an unusual egg of her own instead of a cuckoo's egg, even at a nest that has not been parasitized. Even without errors, hosts risk damaging their own eggs when trying to puncture cuckoo eggs. These costs and errors reduce the benefits of egg rejection and select against it. Therefore, the observed rejection rate at the population level may reflect an evolutionary equilibrium between these opposing forces.

An error in rejection. This great reed warbler is removing her own egg instead of the cuckoo egg – a costly mistake!

For an adaptation to increase in frequency through natural selection, the benefits must be higher than the costs. If we look at various European passerines we can easily see how interactions with cuckoos have shaped some of their traits. We find considerable variation in how good they are at recognizing adult cuckoos or cuckoo eggs. Species with specific food or nest site requirements that are incompatible with those of the cuckoo often show poor defences against cuckoos. For example, seed-eating finches and hole nesters, such as tits and flycatchers, do not in general show any specific aggression towards cuckoos and do not reject foreign eggs added to their nests in experiments. There has been no selection for the evolution of host defences in these species because cuckoos rarely use them as hosts and are unsuccessful if they do.

At the other end of the scale we find suitable hosts that show exquisite egg rejection abilities and specific aggression towards cuckoos. Some, such as bramblings, whitethroats and great reed warblers, are current favourites of cuckoos, while others, such as blackcaps, yellowhammers, chiffchaffs and red-backed shrikes, are apparently not used as hosts at present, at least over most of their range, although they probably were regularly parasitized by cuckoos in the past. We know for example that red-backed shrikes were heavily parasitized in Central Europe only a few decades ago, but have evolved such good defences that cuckoos formerly utilizing them have not been able to keep up and have apparently abandoned them as hosts.

So far we have described coevolution between a host and parasite as if the process would involve all members of the species. This simplification is misleading – in reality, only some host populations may be parasitized at any moment (in coevolutionary hotspots) whereas other populations of the same species do not encounter cuckoos (in coevolutionary coldspots). Some populations are parasitized so heavily that they cannot maintain themselves (sink populations) and would die out were it not for immigration from populations in which parasitism is absent or less severe (source populations). Such differences in cuckoo parasitism pressure between host populations apply not only in space, but also in time as parasitism rates fluctuate from year to year. Previously cuckoo-free areas may become newly 'infected' and formerly parasitized host populations are 'liberated' when cuckoos become locally extinct or switch to different hosts. At a species level the picture is dynamic as host and parasite populations change from year to year and from place to place, forming what biologists call 'metapopulations' – a group of populations of the same species that are separated in space but which interact as individuals move from one population to another. Obviously, anti-cuckoo adaptations are only selected for where cuckoos are a threat. But if individuals wander from their place of hatching then populations will mix and their genes will flow between cuckoo-infested and cuckoo-free populations, altering the prevalence of local anti-parasite adaptations. Such gene mixing is a major reason for the observed geographic variation in host defences and why host defences are never perfect in any one place.

12. One flight over the cuckoo-land – A gallery of deceivers

Throughout this book we have examined how evolution through natural selection has resulted in an array of adaptations in both cuckoos and their hosts, so that cuckoos are better able to trick their hosts, and hosts are better able to avoid being tricked.

But the common cuckoo is only one of approximately 100 obligate avian brood parasites, only about a half of which are cuckoo species. Some of the adaptations of common cuckoos are also found in some other parasitic species, but studies of brood parasites overall have revealed novel strategies not used by our cuckoo. Here we will look at a small selection of other species of cuckoos to show how natural selection can work in different ways to create a variety of strategies, all focussed on achieving the same goals – improved trickery and trickery avoidance.

The mafia strategy – you had better take good care of my kids

The chicks of some brood parasites, such as the great spotted cuckoo in Southern Europe, do not evict the host chicks from the nest but grow up with them instead. Raising one's own chicks as well as a young cuckoo is tremendously costly, and magpie nests containing both magpie chicks and a cuckoo chick often

Great spotted cuckoo chicks do not evict the host eggs or chicks and grow up with their foster siblings, a hooded crow chick in this instance. Their faster hatching (incubation lasts 13 days in great spotted cuckoo versus 19 days in hooded crow) provides a major developmental advantage (Photo: Yoram Shpirer).

fledge fewer magpies than those where no cuckoo is present. It would therefore be beneficial for magpies to reject either the cuckoo egg or chick. Studies have shown, however, that cuckoos may punish those hosts that dare to remove their progeny – the cuckoo parents actually retaliate by destroying the eggs and killing the chicks of any host that dares to reject either their eggs or young. This is an extremely severe penalty and probably worse than just meekly accepting the cuckoo chick in the nest. The hosts will have to work hard but will be able to rear at least some of their own chicks. The cuckoos make their intentions clear by periodically visiting their parasitized nests to check up on how the foster parents are treating their offspring and so force them to care for their unwelcome guest.

Wing patches and wing shaking – can't you see I am hungry?

Chicks of the Horsfield's hawk-cuckoo from Eastern Asia resemble our common cuckoo in that they evict all of the host young from the nest. If there are fewer nestlings to beg for food reduced stimulus can cause the feeding rate of the foster parents to fall, presenting a challenge to the cuckoo which needs to grow to a large size – and as quickly as possible. Several years ago researchers studying the Horsfield's hawk-cuckoo discovered an amazing adaptation to overcome this problem. They noticed that when the foster parents came to feed the cuckoo chick, it raised and shook one of its wings. So far, a normal behaviour similar to that of the common cuckoo. But astonishingly, there is a yellow, featherless patch under the chick's wing that resembles the gape of a young bird – the hosts are fooled into believing that there are several chicks in the nest and may even try to feed the 'fictive chick', the underwing patch. Experiments showed that the yellow wing patch led to increasing provisioning rates for the cuckoo chick. This 'silent begging' may also be preferable to the loud begging calls of our cuckoo because it may attract less attention from potential predators.

How many chicks are in this nest? *Horsfield's hawk cuckoo chicks not only have a yellow gape, but also expose their yellow wing patches to create the illusion of several young, so prompting their foster parents (here, a red-flanked bluetail) to increase their feeding rate* (Photo: Keita Tanaka).

Getting rid of competitors – the murderous chick

It may seem cruel to us when a common cuckoo chick evicts the eggs and young from the nest, one by one. However, among other parasites we can find even more merciless strategies to rid of themselves of competition from host young. American striped cuckoos parasitize hosts that breed in domed nests, making it difficult for their chicks to evict the host eggs and young. Striped cuckoo chicks are equipped with raptor-like hooks on their bills, which they use to stab their foster siblings to death. The dead chicks are then removed by the foster parents, as they would remove chicks dying normally from other causes. The same strategy is used by parasitic African honeyguides, which are not members of the cuckoo family, as a result of convergent evolution (that is, where the same strategy evolves independently in unrelated species).

Australian cuckoos and their hosts – complex strategies Down Under

Research on bronze-cuckoos and their hosts in Australia have revealed some fascinating adaptations that, so far, have not been found anywhere else in the world. Australia has three species of bronze-cuckoos, each specializing in a different group of hosts, all of which build domed nests or nest in holes: Horsfield's bronze-

Equipped to kill.
(·) Greater honeyguide chicks are born with razor sharp bill hooks that they use to kill the host chicks. (··) A greater honeyguide chick inside the underground nest chamber of its swallow-tailed bee-eater host with an un-hatched host egg and chicks (Photos: Claire Spottiswoode).

cuckoos mainly parasitize fairy-wrens; shining bronze-cuckoos usually choose thornbills; and little bronze-cuckoos use gerygones (small superficially warbler-like birds) as their main hosts. Horsfield's bronze-cuckoos lay eggs that closely resemble those of their main hosts, while the other two apparently lay dark brown or olive non-mimetic eggs, which makes them very hard to spot inside the dark nests. These eggs are therefore cryptic, not mimetic, and are always accepted by the host. It seems that the cuckoos really have the upper hand – how could fairy-wrens and other hosts counter such trickery?

The bronze-cuckoo chick evicts the host eggs and young soon after hatching, but fairy-wrens and gerygones often strike back by deserting the nest or even throwing cuckoo chicks out of their nests, the latter behaviour being quite unlike anything seen among European cuckoo hosts. To defend against this behaviour, bronze-cuckoos have evolved chick mimicry – each of the bronze-cuckoos producing young which match the chicks of their favourite hosts both in colour and voice.

But the arms race continues and a recent study has shown that some hosts can now detect even mimetic cuckoo chicks. Superb fairy-wrens call to their eggs during incubation, and on hatching the young fairy-wrens can produce similar calls. Cross-fostering experiments have revealed that the chicks' calls are not genetically determined but instead learned from their mothers while the chicks were still in the egg. Chicks experimentally placed into a foster mother's nest and incubated by her reproduce her calls and not those of their genetic mothers. Fairy-wren mothers respond more often to the calls of young that have learned their calls during incubation and so, in natural situations, can recognise the alien calls given by chicks that are not their own – including bronze-cuckoos which do not learn the calls – and so tell if they have been the victim of a parasitic cuckoo attack!

13. Behind the scenes – How is cuckoo research done?

Imagine you want to learn about the private life of a common bird, such as the great tit. You can put up nest-boxes in a nearby forest, wait for the new tenants to move in and then simply observe them in their own 'living room'. Learning about the private life of cuckoos is far less straightforward. Adult cuckoos, especially females, are professional skulkers – you can easily spend several days hearing cuckoos from dawn till dusk, and even dusk till dawn, yet not glimpse a single cuckoo feather. Cuckoo eggs and chicks are rare and even in host populations described as 'often parasitized', over 95% of the nests are cuckoo-free. It is no surprise that most of the recent novel information on cuckoos comes from a handful of small sites where cuckoo densities are locally high: flooded ditches in Wicken Fen in Cambridgeshire, UK; Apaj Puszta in Hungary; fish-ponds in Lednice and Lužice in South Moravia; the Danube flood plains in Bulgaria; and pine forests in Finland.

Finding nests for a research project. Bård Stokke systematically searches for marsh warbler nests in north-western Bulgaria (Photo: Frode Fossøy).

In this chapter we will briefly introduce you to some of the methods that scientists use to study cuckoos, techniques without which this book could not have been written.

Eggs

Our current detailed knowledge of the brood parasitic behaviour of the cuckoo is the result of two centuries of scientific endeavour, initially carried out by amateurs, but today increasingly the domain of professional research scientists. In fact cuckoo research is a classic example of this trend in the history of science. Crucial facts in the story arose from a branch of ornithology that for many years now has been thought disreputable, and which is today even to some extent illegal – oology, or the scientific study of birds' eggs. Scientific research is subject to fashion and although it is now unthinkable, 100 years ago oology was very much in vogue. The great majority of ornithologists at that time possessed an egg collection as a matter of course, just as botanists had a herbarium or entomologists a case of pinned butterflies. In oology, scientific interests combined with the passion for collecting, though the latter was usually given first priority. It was the similarity in appearance of the eggs of host and parasite that first aroused the interest of egg collectors and ruthless competition between rival collectors drove them to ever greater efforts to amass large collections of rare and unusual eggs from rare and unusual species.

If Edward Jenner answered the first big question raised by the breeding biology of the cuckoo (how does a cuckoo chick evict the contents of a host nest?) the second – how does the cuckoo's egg get into the host's nest? – was answered by the Birmingham

Drawers full of eggs. Egg collections, such as the Čapek collection at the Moravian Museum at the Budišov Castle in the Czech Republic, provide important sources of information for research.

manufacturer Edgar Chance (1882-1955) although not until 130 years after Jenner's work! Chance, was an avid egg collector, who was continually on the look-out for eggs of rare birds to add to his collection, an addiction that eventually resulted in his expulsion from the prestigious British Ornithologists' Union. He had not only a unique talent for finding nests but also a lively and innovative mind. His perfect study area was a lucky find, a 20 hectare heath, used as a common, surrounded by woodland and containing 6–10 breeding pairs of meadow pipits that were parasitized by one or two female cuckoos every year. Together with his helpers he systematically searched for every pipit nest over a seven year period.

He manipulated the pipits by collecting their eggs, so that they continually laid replacement clutches, presenting the cuckoos with ever more parasitic opportunities. In this way he caused 'female cuckoo A', a summer visitor of several years standing, to lay 25 eggs in a single season – a performance worthy of the Guinness Book of Records, considering that female cuckoos average only around a third of this 'clutch-size'. In his 1922 book, *The Cuckoo's Secret*, he proudly wrote: 'We have been able ... to engineer circumstances ... and acquired such a control over the actions of a cuckoo, that, on many occasions, after prophesying the day and hour when she would lay, and the very nest in which she would deposit her egg, we have seen the forecast accurately fulfilled'. He finally brought his 'art' to such a peak of refinement that he invited spectators to watch the spectacle from a hide near the predicted nests. With Oliver Pike he succeeded in capturing the entire process surrounding the laying of an egg in a meadow pipit's nest on film – one of the first natural history movies ever made. In 1921 this was a ground-breaking achievement that excited great attention and its first London audiences found it simply breath-taking.

Despite their great difference in age, Chance was in contact with three other famous collectors of cuckoo's eggs, the natural history dealer Eugene Rey (1838-1909) from Leipzig, the schoolteacher Václav Čapek (1862-1926) from Oslavany (near Brno, Czech Republic), and E. C. Stuart Baker (1864-1944), the then Inspector-General of the Assam Police. Rey's collection contained hundreds of cuckoo eggs. He had constructed an apparatus to measure the thickness of the shells of birds' eggs because the shell of the cuckoo egg seemed to him to be so much thicker than that of its hosts. Over a period of precisely 30 years, the dedicated Čapek had found exactly 1500 parasitized clutches. Tragically, he had to sell large parts of his valuable collection to raise the money needed to compensate the family of a boy whom he had accidentally injured during a hunting party. Perhaps for this reason he later turned to palaeontology, making a second career for himself as a specialist in fossil birds. But Rey, Čapek, and Chance were trumped by Stuart Baker, and his collection of 6000 cuckoo eggs (belonging to a variety of cuckoo species) – as an Inspector-General, Baker commanded a small 'army' of well-trained nest finders. All four published their studies in books and papers, making great advances in the discovery of the 'cuckoo's secrets' using the methods of their time.

Even if these oologists and their many colleagues were rather excessive in their egg-collecting activities, it is fortunate that their often very carefully documented collections are accessible to us in natural history museums today, where they can be studied by new generations of researchers. The egg collectors of the past could never have imagined that their material would one day become the foundation of so many brood parasitism studies that would never have been possible otherwise.

Many thousands of cuckoo eggs are held by museums all over Europe, most of them collected between about 1850–1930. These collections provide unique opportunities to study changes in traits such as egg mimicry, in space and time; to investigate differences in egg colours and markings between different cuckoo tribes; and to investigate geographical variation in host use by cuckoos.

Some of the early pioneers such as Bernhard Rensch acted like cuckoos themselves and performed simple experiments adding eggs to the nests of various passerines and observing if the eggs were accepted or rejected. Experiments of this type remain important and are widely used by scientists today.

How do we assess egg mimicry? One traditional approach is to photograph the eggs and then score the degree of similarity using a predefined scale. Of course, this is a rather subjective method depending on the skills and judgement of the particular observer, so more objective methods have been developed. For example, computer software has been written to analyse images, and spectrophotometry is used to objectively measure the colour reflectance from the egg surface. Modern computational methods in 'visual modelling' can account for most aspects of avian vision and properties of the environment that ultimately determine what a bird actually sees.

Securing DNA from old eggs. New techniques have made it possible to obtain DNA from eggs that have been stored in museums for over 100 years.

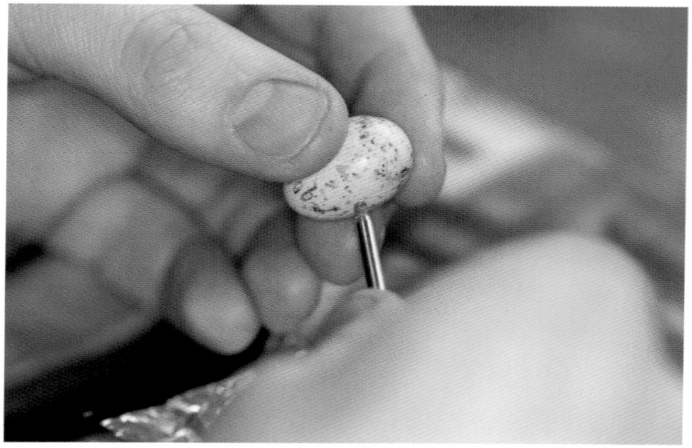

All these methods have their pros and cons. While the image analysis method is subjective and does not consider UV reflectance – invisible to a human observer but visible to birds – it does have the important advantage that it provides a good view of the whole egg surface with all its patterns and signatures. Spectrophotometry on the other hand is objective and does include UV, but only measures the colour reflectance of a tiny proportion of the egg and misses the larger patterns and signatures. Most recent studies prefer to use a combination of these methods and new techniques are constantly being developed to assess host-parasite egg similarity in more holistic and realistic ways.

When doing research in the field, we try to behave like cuckoos by finding nests and parasitizing them with dummy cuckoo eggs made of plaster of Paris or clay. We can then visit the nests on subsequent days, to see if the hosts have accepted or rejected our dummy egg. We can investigate the ability of hosts to recognize foreign eggs by painting the dummy eggs in various colours and patterns, or by altering the size or shape of the eggs.

Video cameras are indispensable aids in studying host behaviour after planting our eggs. Host reactions can be so fast that it is hard to tell what has happened using the naked eye. Hosts sometimes respond shortly after the nest contents have been manipulated but in many cases, the response is delayed, even for several

Egg rejection experiments. Inserting experimental eggs into nests of various species is an important means of investigating host anti-parasite adaptations. Here, a blue model egg has been placed into a great tit nest (Photo: Tomáš Grim).

days – so using cameras saves precious time that the researcher would otherwise need to spend watching the nest from a hide.

Nestlings

Naturally, it is harder to study cuckoo chicks than eggs. Eggs are immobile – they do not move and do not try to escape the research worker. Moreover, eggs can be manipulated easily, while chicks allow for very little manipulation – their gapes and perhaps skin can be painted with non-toxic colours, and that's about it.

Chick vocalizations are easier to manipulate than their physical appearance. A tiny loudspeaker can be hidden beside the nest and researchers can play back chick begging calls, or parental alarm calls. In each experiment scientists ask how the treatment may affect the birds' behaviour and what the outcomes might be. We measure various parameters, like chick weight or wing length – the latter being a crucial indicator of likely survival during the post-fledging period when predators take a heavy toll on chicks.

We try to discover whether hosts recognize their own versus foreign chicks by cross-fostering experiments – exchanging the parents' chicks for others of the same species or a different one. By removing older original chicks and introducing new, younger chicks we can effectively prolong the period required for the brood to be raised to fledging. It is then up to parents whether they accept the extra workload or discriminate against these prolonged conspecific broods – and perhaps also cuckoo broods – under natural conditions. Finally, we can easily vary the brood size by adding or removing chicks. In combination with brood age manipulations, this allows us to test various ideas about cues that foster parents use to adjust their parental care – or misplaced care of cuckoos. Simultaneous presentation of their own nestlings and a cuckoo chick side by side allows us to ask whether hosts prefer their own kind or a supernormal alien.

Adults

We can record both host and cuckoo behaviour using cameras, and what great advantages that brings! Cuckoo laying or predatory visits are extremely brief, just a couple of seconds, making unaided visual observations unreliable, or even impossible in cases where the host nests cannot be seen (e.g. in nests placed inside tree or ground holes). Without miniature cameras cunningly hidden inside custom-designed nest-boxes it would be impossible to discover whether common redstart cuckoos bring their eggs to the host nest inside their bill or crop, or lay them directly into the nests of this strictly cavity-nesting cuckoo host.

When does the cuckoo chick's ejection instinct cease? Experiments where an egg is added to a reed warbler nest containing a cuckoo chick helps answering this question. Most cuckoo chicks evict the host eggs when they are 2–4 days old.

However, today's researchers do not rely only on undisturbed observations of natural cuckoo-host encounters at the host's nest. Parasitism rates are typically low and the number of nests available to a cuckoo on a particular day might be high, for example, where reed warblers are breeding at high densities. It is impractical to summon enough observers to watch them all and such logistic constraints favour the use of non-living cuckoo substitutes – ideally stuffed cuckoo dummies or model replicas made from polystyrene or balsa wood. Models may be less than perfect replicas and one might think that hosts would have no difficulty distinguishing them from real cuckoos. Yet birds, including cuckoo hosts, often respond even to very crude and simplified cues provided that the key features are represented, if only in caricature form. Hosts respond to stuffed dummies exactly as to a live cuckoo visiting their nest – they mob them making diving flights from above, giving alarm calls, and even making physical attacks targeting the eyes and nape.

Another unique advantage of dummies is that researchers can specifically manipulate parts of the dummy to see what specific cues trigger a host reaction. If hosts attack a dummy cuckoo with its eyes painted black less often or less intensely than an unaltered dummy with naturally yellow eyes, we can conclude that yellow eyes are an important cue prompting host aggression. By manipulating other supposed cues, such as barred underparts, and combinations of them and other factors such as the distance between the dummy and the nest under investigation, we can assess the relative importance of a variety of cues.

Female cuckoos never call during visits to host nests – only after departing do they utter a peculiar bubbling 'victory call'. It would make no sense therefore to accompany a dummy presentation with a cuckoo call playback. However, it is interesting to play back calls in the general area surrounding host nests during the whole breeding season to fool the hosts into 'believing' that the risks of cuckoo parasitism are higher than they actually are. Researchers can then check how this elevated perceived risk of parasitism affects host responses to both dummy cuckoos (do hosts attack them more vigorously?) and model eggs placed in the nest (do hosts reject them more often?). This type of experiment is important in understanding the plasticity (or rigidity, for that matter) of host anti-cuckoo strategies.

Fooled by a model. Experiments with model cuckoos (dummies) allow researchers to observe host aggressive behaviour towards cuckoos. Here, a hand-held dummy cuckoo is attacked by a great reed warbler.

These sorts of experiments are just the first steps. How do cuckoos behave in the breeding area, and at what time? How large are their territories? When and where do they migrate after the breeding season? Where do they spend their winter? Where do the chicks of the year return to breed in the following season? These are questions that are now under investigation using modern technology such as radio telemetry and satellite transmitters, attaching the equipment to cuckoos in a backpack. These two methods differ in the scale over which they operate. Radio telemetry is used to track cuckoos over small distances in their breeding areas as the operator literally follows the bird using a radio receiver to pick up the signals emitted from its backpack radio transmitter. The technique is used for small scale movement and behaviour studies and meticulous tracking enables us to map the whereabouts of a cuckoo, where and when the female lays her eggs, and so on. Satellite transmitters, on the other hand, operate over continental scales, do not demand such active tracking and effort on the ground, and are used mainly to study migration. The transmitters are in contact with satellites in space and signal their position at specific intervals, no matter where in the World the cuckoo may be. Scientists can therefore sit in front of their computers and download the movements of the cuckoo under study all year around.

We still know very little about the migration pattern of young cuckoos – despite having ringed some 20,000 cuckoos across Europe, there have only been two records of a young cuckoo found in Africa! Such observations only provide information on the beginning and end of the trip, but do not tell us anything about the direction of travel, the routes taken, or the stop-over sites used. The very first information on these fascinating aspects of the long-distance travel of young cuckoo's was obtained by satellite telemetry only as recently as 2014. Furthermore, our knowledge of where young cuckoos return to breed and how they select which host species to parasitize is also very poor. In particular, we still have no observations to confirm one of the most important assumptions behind the whole cuckoo story – we still do not know whether a cuckoo chick hatched in the nest of a particular host really does return from its wintering grounds to parasitize the very same host species!

Under surveillance. Nest recording by video cameras is an important tool in studying brood parasitism.

New ways to peer into the cuckoo's Pandora's box

The cuckoo genome has recently been sequenced and this represents a huge leap forward – new genetic data will allow us to finally ask the most fundamental questions about cuckoo biology such as what is the genetic basis for cuckoo egg colour and how did the various host-specific cuckoo tribes arise? Understanding the details of the cuckoo genome should also open the way to answering other key questions such as how are the cuckoo tribes maintained and how differentiated are they? Molecular techniques bring so many exciting possibilities and will surely allow us to resolve many more secrets of the cuckoo's world in the years to come.

Cuckoos in Finland, which parasitize common redstarts in woodlands, face a completely different habitat compared to their central European reed warbler counterparts.

This cuckoo chick has managed to evict all of the redstart eggs in its adopted nest, a feat not always possible in nests in holes or nest boxes.

•
Acrobatic feeding. A young cuckoo, about to fledge the birdhouse entrance, forces a male common redstart to feed it while hovering.

••
On the edge. A cuckoo fledgling prepares to leave the nest box.

•••
Into the wild. The first flight.

Everything is a new experience. This juvenile cuckoo will be cared for by its common redstart foster parents for another three weeks or so.

pages 134–145
A wide array of foster parents.
Marsh warbler

Sedge warbler

White wagtail

• Robin

•• Red-backed shrike

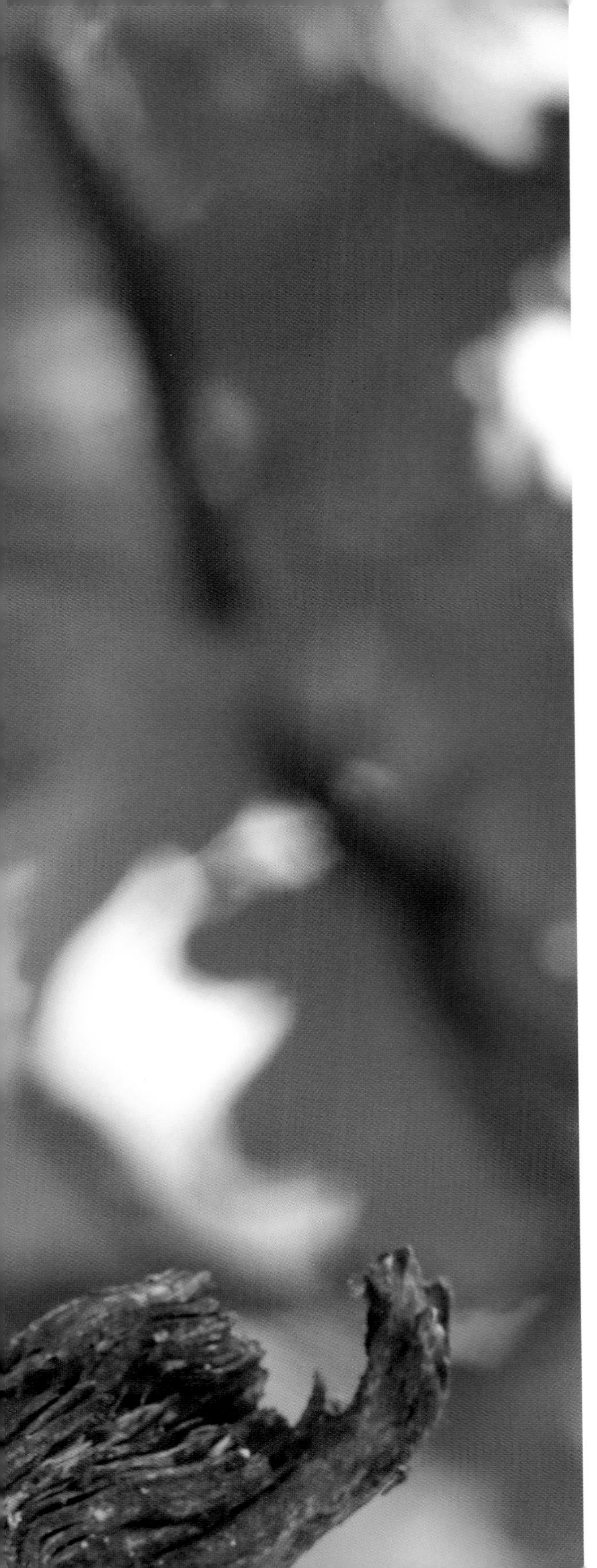

• Common redstart

•• Black redstart

•
Spotted flycatcher

••
Barred warbler

•••
Wren

Cuckoos in a changing world

14. Flying into a precarious future – The decline of cuckoos

When did you last hear a cuckoo? Many European cuckoo populations have declined drastically over the last 20 to 30 years. The species first appeared on the Red List of threatened bird species in several European countries some time ago, although it was once common throughout its entire range. Between 1850 and 1900, when German foresters were combating caterpillar plagues, such as pine looper and gypsy moth, they were able to count on the help of the cuckoo. Sometimes over a hundred greedy cuckoos could be seen gathered in a plantation where the density of caterpillars was at its highest and the stomach of one gluttonous bird contained a total of 173 hairy caterpillars. In today's intensively cultivated European landscapes caterpillar outbreaks are a thing of the past and cuckoos have suffered, though in Eastern Europe cuckoo population densities have remained more stable. Cuckoos also face threats of a completely different kind. In a shocking example from 2015, conservation activists campaigning against bird trapping along the coasts of the Mediterranean were speechless when they discovered large crates packed with cuckoos in Egyptian markets, alongside many other species of migratory birds, all carefully sorted by species. All of them were alive, all had broken wings, and all had come from breeding grounds in Eastern Europe, where extensive agriculture is still widely practised and cuckoos are common, for the moment at least.

The cuckoo is actually a successful species and compared with many others its breeding range is enormous, encompassing the greater part of two continents, Asia and Europe. Although every individual female cuckoo is host-specific, parasitizing only a single species, when it comes to habitat it is a generalist. Cuckoos generally prefer semi-open landscapes and open woodlands when they are searching for food, but can easily use any habitat in pursuit of their own host species – reedbeds for reed warblers, woodlands for robins, heaths and moors for meadow pipits, and certain parts of the Alps for black redstarts. Cuckoo numbers in central Europe have remained at a higher level than those of, for example, golden oriole, hoopoe, wryneck, nightjar and roller – five species comparable in size, diet, and migratory habits to the cuckoo but different in that their habitat requirements are much more specialized.

It therefore sends a very worrying message about the health of the environment that the central and western European populations of even this 'robust' generalist have declined so rapidly since the 1980s. Cuckoos have almost disappeared from some western parts of their range. For example, the British Breeding Bird Survey has recorded a decrease of up to 85% between 1994 and 2015. But why has this brood parasite suffered such losses? Could the 'cuckoo clock' perhaps be awry because of a fall in the numbers of host species, or because climate change has knocked the breeding seasons of parasite and hosts out of synch? Do cuckoos face increasing problems on their migration routes and in their winter quarters in the heart of Africa, or is their food supply in decline? We now know so much about the cuckoo as a brood parasite, but next to nothing about the reasons why the species is declining so rapidly.

The fall in cuckoo numbers in Britain was thoroughly described by Chloe Dennerly in her unpublished research thesis in 2014. It has not been due to a decline in the numbers of host species – indeed populations of the commonest hosts, such as dunnocks, meadow pipits, and reed warblers have hardly declined, and in some areas have even increased. Although she remarked on the slight mismatch observed in the breeding cycles of early-breeding host species and cuckoos due to climate change, this effect was too small to account for such a rapid decrease in cuckoo abundance. The decline in numbers is markedly stronger in the south of Britain than in the north, so could this latitudinal trend arise from differences in the migration routes of different populations? Cuckoos breeding in England migrate on a south-westerly route while those in Scotland head in a more south-easterly direction. It has been known for some time that, in general, west-heading migrants suffer increasingly greater problems on their migration routes – as well as on their wintering grounds – than do east-heading conspecifics, for a variety of reasons including the increasingly drier late summer conditions, and increasing agricultural intensification and other human related changes in land use in southwestern Europe. Populations of the former are therefore declining more rapidly than those of the latter. But if this were the only reason we would expect the decline in English cuckoo numbers to be uniform across different habitats, but this is not the case. The decline is stronger where land use is most intensive, suggesting a direct connection with the intensification of agricultural practice. England has much more farmland than Scotland, placing greater pressure on cuckoo populations in the south than in the north. Wherever semi-natural habitats – unimproved or less fertilized grasslands, heaths, and moors – remain, cuckoo numbers have not declined and there are many more such areas in Scotland than in England. In some parts of north-west Scotland the cuckoo population has even increased, by as much as five-fold in places where moorland has been planted with conifers. The young pines provide the cuckoos with vantage points from which to search for meadow pipit nests, which in the previously open, treeless habitat were almost impossible to spot. Whether these advantages will persist as the plantations mature remains to be seen.

As agricultural land becomes increasingly intensively cultivated, it loses quality as a habitat for many plants and animals and its species richness declines, as does the number of individuals of all species. This is particularly noticeable for butterflies and moths, and the macromoth species whose hairy caterpillars provide cuckoos with their main food supply have suffered an especially marked decline. It appears that the reduced availability of cuckoo prey is a strong driver of the species' decline in agricultural ecosystems. Additional factors such as the 'tidying-up' of the landscape and falling numbers of some host species produce a cumulative effect, accelerating the decline in cuckoo abundance.

Monocultures are deserts for biodiversity. **Intensive agricultural land is avoided by both cuckoos and their hosts** (Photo: Klaus Leidorf).

If the food supply of adult cuckoos is reduced then they must invest more time in foraging, time that becomes unavailable for monitoring hosts and their nest-building activities. This then reduces the number of eggs they can lay and fewer offspring are produced as a result. The decline in cuckoo populations in parts of England with intensifying agriculture, where the dunnock is the principal host species, could possibly lead to the extinction of the dunnock cuckoo tribe.

A typical area of artificially created fish ponds. **Such habitats are optimal for cuckoos and their sedge-, marsh-, reed- and great reed warbler hosts** (Photo: Klaus Leidorf).

Soon to be only a memory? A juvenile cuckoo flies into a precarious future. In many areas, cuckoos are in heavy decline.

So this extinction would not result from the hosts having evolved better responses to reject the parasite, but from man-made alterations to the landscape on a historically unprecedented scale. There may still be sufficient hosts for the cuckoos, but not enough food, a shift of this sort probably never having occurred before. Both the geographical distribution and the habitats and host spectrum of the cuckoo in Western Europe have changed rapidly within a mere 30 years as agricultural intensification has increased.

This book has been about the cuckoo as a brood parasite and about its interactions with its host species. We now begin to realize that it is not just about the participants in this story, but also about the major role increasingly played by humans and our management of the land. Mankind has always created and destroyed habitats. Witness the many fishponds in central Europe, whose reedbeds provide a habitat for both reed warblers and reed warbler cuckoos. But too often, old ponds have been filled in and new ones dug elsewhere, creating a lasting impact on the parasite-host relationship. For example, narrow, linear reedbeds lined with trees do not historically border natural lakes, but are instead typical of man-made ponds and facilitate cuckoo parasitism. Mankind has altered the natural environment through cultivation for thousands of years, but the problem of our time is simply the scale of our current management – we exploit the land more and more intensively and the rate of landscape change is rapidly accelerating. Modern agricultural practices transform structurally rich and biologically diverse habitats, degrading them into uniform monocultures.

Until now, large-scale intensive agriculture has had a negative impact on habitat specialists, such as the grey partridge, corn bunting, and skylark. The common cuckoo is the first habitat generalist whose populations now face massive threats, sending a clear alarm call to conservationists.

Given the resulting state of affairs, how can we help this charismatic bird? One question opens the door to yet another set of questions. Do we have to treat each cuckoo tribe as a distinct subspecies needing individual protection and if so, which host is the most profitable for the cuckoo? Should we provide help in small increments or on a grand scale? If we want to help conserve the generalist cuckoo we must save and protect most of the habitats it uses, thereby also conserving all of the biological diversity in those environments. Considering the complexity of the ecological web within which this parasite has its niche – and taking its specific host and prey species into account – this is without question the most promising, and challenging, approach. The answers to these questions can only come from further conservation biological research, using telemetric studies, diet analyses, and much more.

An obvious priority should be to safeguard the habitats where cuckoos still occur. In Western Europe this would be semi-natural landscapes. In the face of the gigantic area covered by industrialized agriculture and the huge loss of biodiversity in the cultivated landscapes of Europe, completely new approaches will be required, a truly daunting task. A network of large-scale agrarian Biosphere Reserves would need to be created as ecological compensation areas, not only to help the cuckoo but to promote habitat biodiversity in general. A simple, but very effective, first step would be a shift towards organic farming. New reedbeds could also be created very economically, and are always attractive to cuckoos given their high reed warbler nest densities. But above all, we must ensure that the same mistakes that have been made in the west are not repeated in Eastern Europe, where extensive agricultural methods are still widespread over large areas.

The dictum 'act locally, think globally' requires that we also have to protect those regions through which cuckoos pass on migration as well as their wintering grounds – a gargantuan task indeed. However, the most important thing is that everyone learns to appreciate and value the natural world around us on which we depend, as well as to understand the root causes of the environmental problems facing us on our overcrowded planet. This, the most crucial task of nature conservation can be achieved by promoting public awareness and education. Then perhaps, it will not come to the worst – that we say farewell forever to the Harbinger of Spring.

Lost. Robin eggs evicted by a young cuckoo.

Passion for a parasite

15. Photographing the secret life of the cuckoo

It is rare to find a combination of circumstances so conducive to taking the sort of photographs that appear in this book, and many readers will no doubt be interested in the background story. Since early childhood, the photographer Oldřich (Olda) Mikulica has lived in the southern Moravian village of Lužice, Czech Republic, not far from Hodonín, the district's largest town. Lužice lies on the edge of a large area of fishponds where fish, mainly carp, have been raised for the table for centuries, and Olda's home is right beside one of them. This multitude of ponds is a paradise for a great many bird species that nest at the water's edge, and their extensive reedbeds are home to high breeding densities of great reed warblers and reed warblers, as well as marsh and sedge warblers. These are perfect conditions for a brood parasite, so cuckoos have always been part of Olda's daily life both as an amateur ornithologist and nature photographer. Even as a child he watched the bizarre scenes as young cuckoos grew in the nests of reed warblers and wagtails. So it comes as no surprise that he has been fascinated by the world of this parasite and its hosts for the last 35 years. In this time Olda has peered into hundreds of songbird nests containing cuckoo eggs or young, and knows the biology of these avowed opponents to the smallest detail.

The photographer. Portrait of Oldřich Mikulica (Olda) at work.

Waiting patiently. *Olda's hide in the fish pond area, in front of a great reed warbler nest. Hour after hour is devoted to sitting quietly inside, waiting for the right moment for a photo.*

This lucky constellation of circumstances has captivated not only Olda. Researchers from the Institute of Vertebrate Biology, The Czech Academy of Sciences, Brno, Czech Republic, have also been working in the Lužice fishponds for over 20 years, studying among other things the interaction of the cuckoo and its hosts, often in collaboration with evolutionary biologists from Trondheim University, Norway, and their work has resulted in many exciting discoveries.

Most of the photos in this book were taken at the Lužice ponds, with a few others from central Hungary and southern Finland. Photographing the secret life of

Devotion. Jana's dedication to her cuckoo-fascinated father Olda.

the cuckoo requires incredible patience, preparation, and experience, as well as a sizeable helping of good luck. Everything has to come together for the perfect moment – weather, light, and the nest surroundings. The photographer has to predict when the female cuckoo is going to lay and in which nest – it's all about being in the right place at the right time. The egg is laid into the host nest in a few seconds, and Olda has captured that moment time and again. This means he has to search for many nests, observe them from the nest-building phase onwards, and precisely learn the

The next generation. Olda's grandson Viktor holding a ringed cuckoo chick in his caring hands.

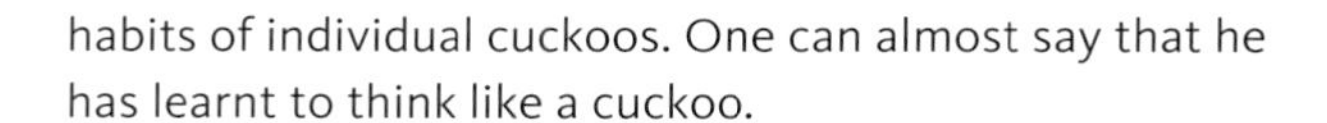

habits of individual cuckoos. One can almost say that he has learnt to think like a cuckoo.

Olda has spent the equivalent of many months crouched inside a camouflaged tent and when spring arrives there is only one subject of household discussion – cuckoos. Given such intense dedication, some cuckoos and reed warblers have become his personal friends. Over the years Olda has witnessed many strange events. For example, in 2016 his 'brown girlfriend' (see photos on pages 68 and 69) had to suffer numerous beatings from great reed warblers, but still managed to lay her egg in one of their nests. Olda checked the egg during the afternoon after it had been laid. Next morning the egg in the nest was most definitely a different one! The automatic video camera that had been installed at the nest by researchers from Brno confirmed what had happened – after Olda had left the previous day, a second female cuckoo appeared in the late evening, removed the first cuckoo egg, and laid her own in the nest. Without continuous documentation, as well as Olda's experience in recognising cuckoo eggs, this would never have been discovered. Sometimes the most incredible stories are told by real life itself!

* The cuckoo female on page 152 is carrying a cuckoo egg in her bill. According to old myths, cuckoos brought their eggs into host nests by the bill. This was not the case here. She simply removed one egg from the host nest during laying, and the stolen egg happened to be an egg from another cuckoo female that had laid in the same nest a few days back (an example of multiple parasitism).

Acknowledgements

Cuckoos are easy to hear but difficult to watch, photograph, and study. However, studying them becomes a lot easier when friends and colleagues cooperate, and we must therefore express our deep gratitude to the many research workers with whom we have exchanged opinions and ideas – in some cases over decades – on the great variety of puzzles and problems in cuckoo biology.

Several text illustrations have been generously provided by close colleagues, and it is a pleasure to thank Mohammed Mostafa Feeroz, Mikkel Willemoes Kristensen, Klaus Leidorf, Bruce E. Lyon, Per-Harald Olsen, James W. Rivers, Yoram Shpirer, Claire Spottiswoode, Keita Tanaka and Kasper Thorup. Valuable help in our search for photographs was provided by Sajeda Begum, Tomás Pérez-Contreras, Mominul Islam Nahid, Brian D. Peer, Manuel Soler and David J. White. We are particularly grateful to Klaus Nigge, who, with the skill and eye of an artist, teased the best out of many of the photos. Douglas G. D. Russell, Curator of the egg collection at the Natural History Museum at Tring, allowed us to take photographs of specimens in his care.

In the Czech Republic and Finland, Oldřich Mikulica and Tomáš Grim were supported in a multitude of ways by Jaroslav Bobčík, Dušan Boucný, Marcel Honza, Ivan Hurbánek, Václav Jelínek, Tomáš Koutný, František Krause, Michal Kysučan, Petr Procházka, Josef Ptáček, Ondřej Ryška, Peter Samaš, Miroslav Šebela, Michal Šulc, Alfréd Trnka, Zdeněk Tyller and Libor Vaicenbacher. Additionally, Tomáš Grim expresses his thanks to the Human Frontier Science Program (RGY69/2007 and RGY83/2012) and the Czech Science Foundation (especially P506/12/2404) for their long-term financial support of his cuckoo studies. Bård G. Stokke acknowledges the support and friendship of his long-standing collaborators Frode Fossøy, Arne Moksnes and Eivin Røskaft at NTNU in Trondheim, Norway.

Some collaborators deserve special mention. Klaus Nigge advised us on all aspects of book illustration. Brian Hillcoat translated the German texts and made many useful suggestions in other areas. Wise and experienced advice from Tim Birkhead encouraged us to continue the project at a difficult time when its continuation appeared at risk. We thank Andrew Richford for his superb copy-editing; he identified so closely with the manuscript that working with him was a genuine pleasure. Elske Verharen of Oxedio has achieved wonderful things with the layout, never losing her patience in the face of our various wishes. Jack Folkers, our publisher at KNNV Uitgeverij in Zeist, made our project his own, guiding it skillfully and constructively through all phases of its production. We consider ourselves greatly honoured that Nick Davies has enriched the book with his obliging and genious foreword.

Our greatest thanks however go to our families, especially to our partners for their apparently limitless patience with these 'cuckoo nerds' over many years. This book would not have been possible without their understanding and goodwill.

Suggested further reading

In writing this book, we have benefited greatly from the brilliant observational and experimental work carried out by many keen naturalists and scientists. For ease of readability, we have chosen not to include in-text citations, but do want to express our gratitude to all of those who have so enriched our understanding of cuckoo biology. For any of our readers who want to learn more about cuckoos and their exciting lives, we suggest a few up to date books in a variety of languages (published post 2000) for further reading.

Davies N. B. 2000: Cuckoos, cowbirds and other cheats. T & AD Poyser, London.

Davies N. B. 2015: Cuckoo: cheating by nature. Bloomsbury Publishing PLC, London.

Ericson P. G. P. & Sjögren H. 2004: Boken om göken. Atlantis, Stockholm. (In Swedish)

Erritzøe J., Mann C. F., Brammer F. P. & Fuller R. A. 2012: Cuckoos of the world. Christopher Helm, London.

Haikola J. & Rutila J. 2008: Käki. Otava, Helsinki. (In Finnish)

Hellebrekers A. W. 2004: Heeft de Koekoek overlevingskansen? Datawyse Boekproducties, Maastricht. (In Dutch)

Isenmann P. 2006: Le coucou gris. Editions Belin, Paris (In French)

Moksnes A., Røskaft E. & Stokke B. G. 2011: Gjøkens forunderlige verden. Tapir Akademisk Forlag, Trondheim. (In Norwegian)

Numerov A. 2003: Interspecific and intraspecific brood parasitism in birds. FGUP IPF Voronezh. (In Russian)

Payne R. B. 2005: The cuckoos. Oxford University Press, Oxford.

Zech J. 2008: Zum Kukuck nochmal. Höma Verlag, Offenbach. (In German)

The authors

Oldřich Mikulica is a nature photographer with a special interest in the common cuckoo and its biology spanning more than 35 years. He was crowned European Nature Photographer of the Year in 2011 and is the author of several illustrated natural history books and well received films (for Czech TV). A retired engineer he lives in Lužice, Czech Republic.

Tomáš Grim is Professor of Zoology at Palacký University, Olomouc, Czech Republic. He has studied cuckoos and other birds for 20 years and his ecological research projects have taken him all over the World. He has written 50 scientific papers on cuckoo biology and has been popularising biology via writing popular articles and giving public lectures since the age of 16.

Karl Schulze-Hagen is a gynecologist and dedicated ornithologist and has studied reed warblers and cuckoos for over 40 years. He is the co-author of *Reed warblers – diversity in a uniform bird family* published by KNNV, Uitgeverij and the winner of the British Birds/British Trust for Ornithology Best Bird Book of the Year 2012. He has written several scientific papers on cuckoo biology and natural history and lives in Mönchengladbach, Germany.

Bård Gunnar Stokke is a researcher in conservation biology and evolutionary ecology at the Norwegian Institute for Nature Research (NINA). He is also holding a post as research scientist in the AfricanBioServices project at the Department of Biology, NTNU in Trondheim, Norway. He has been studying the ecology of avian brood parasites for 20 years and is the co-author of several books and more than 65 scientific papers dealing with brood parasite biology.

Photography Oldřich Mikulica

Text Tomáš Grim, Karl Schulze-Hagen & Bård G. Stokke

Foreword Nick Davies

Concept Karl Schulze-Hagen

Copy-editing Andrew Richford

Design Elske Verharen, Oxédio, www.oxedio.nl

Originally published in The Netherlands in 2016
by KNNV Uitgeverij, Zeist, 2016

Published in the United Kingdom in 2017
by Wild Nature Press
Winson House, Church Road
Plympton St. Maurice, Plymouth PL7 1NH

A CIP catalogue record for this book is available from the British Library.

ISBN 978-0-9955673-0-6

www.wildnaturepress.com

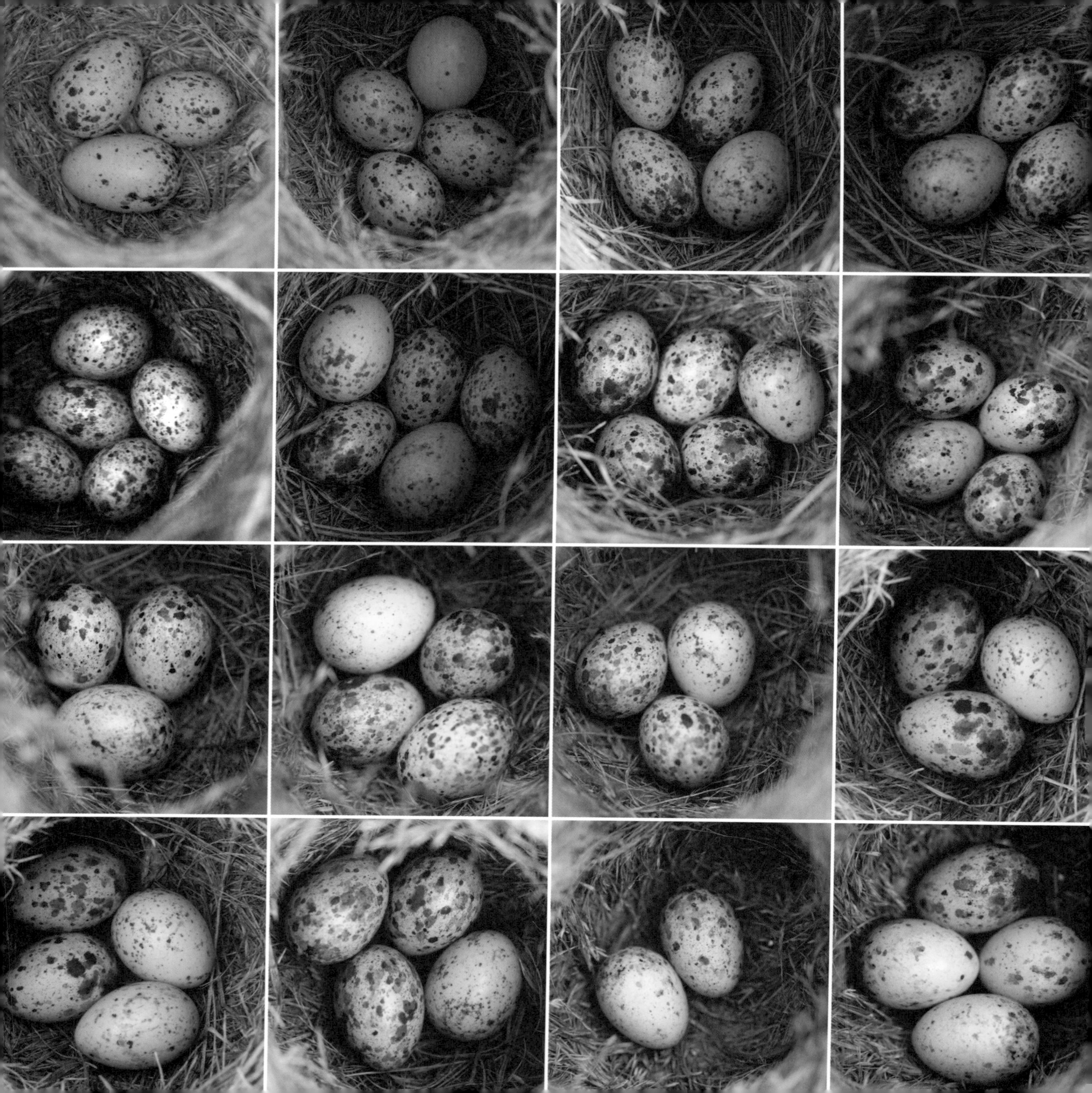